Benjamin Franklin's Numbers

Benjamin Franklin's Numbers

An Unsung
Mathematical Odyssey

Paul C. Pasles

PRINCETON UNIVERSITY PRESS   PRINCETON AND OXFORD

Copyright © 2008 by Princeton University Press

Published by Princeton University Press, 41 William Street, Princeton, New Jersey
08540

In the United Kingdom: Princeton University Press, 3 Market Place, Woodstock,
Oxfordshire OX20 1SY

British Library Cataloging-in-Publication Data is available

Library of Congress Cataloging-in-Publication Data

Pasles, Paul C., 1968–
Benjamin Franklin's numbers : an unsung mathematical odyssey / Paul C. Pasles.
p. cm.
Includes index.
ISBN-13: 978-0-691-12956-3 (hardcover : alk. paper)
ISBN-10: 0-691-12956-8 (hardcover : alk. paper)
1. Franklin, Benjamin, 1706–1790.   2. Mathematics—History.   3. Magic squares.
4. Mathematical recreation.   I. Title.
QA24.P37 2008
510.92—dc22

2006102508

This book has been composed in ITC Garamond Book with
Snell Roundhand Black Display

Printed on acid-free paper. ∞

press.princeton.edu

Printed in the United States of America
1 3 5 7 9 10 8 6 4 2

For Antoinette
who loved history

and Anthony
who loves to count

Contents

Preface

ew lives have been as well examined as that of Benjamin Franklin. New biographies continue to spring forth regularly, with no end in sight. Somehow, though, the mathematical side of that great mind remains unstudied. Hence the present book, *Benjamin Franklin's Numbers*.

Scattered throughout this historical account are various mathematical questions. You may decide to attempt some, all, or none of these exercises, depending on your personal preferences (and your mathematical background); but at least read them, to see what mathematics can do. Their actual solution is not necessary for a proper reading of this book.

Seventeenth-century documents present some peculiar challenges. Often I have updated or corrected spellings, capitalization, and punctuation. In other cases, when it is not too distracting, the original text has been left unmolested, the better to convey some sense of authenticity. I make no claim of consistency on this score.

I have not written anything resembling a complete biography. As with all great lives, Franklin's life appears to us as a magnificent jigsaw puzzle with a few pieces missing. With this book, I hope to fit one more piece firmly in its place.

Benjamin Franklin's Numbers

1   The Book Franklin Never Wrote

It seems to me, that if statesmen had a little more
arithmetic, or were more accustomed to calculation,
wars would be much less frequent.
   —Benjamin Franklin (1787) [1]

The American author Ernest Hemingway never composed
a guide for writers. Indeed, the very idea was anathema to him, in
part because of a superstitious fear that any such discussion of his
art would destroy the thing itself, just as dissecting a flower dis-
solves the very essence of its beauty. Yet there are enough fragments scattered through his private correspondence, in interviews,
and in the opinions of his fictional characters, to piece together exactly what he would have opposed: a book called *Ernest Hemingway
on Writing*.[2] Likewise Benjamin Franklin said little regarding his
magic squares, revealing few results and no methods, but on mathematical matters there is enough surviving material to fill a book on
this unexamined side of Franklin's otherwise meticulously documented life. Hence, the present account of Franklin's mathematical
experiences and his miraculous numerical creations.

There is a danger here that we might simply be indulging an
artist who is working outside his usual field of true expertise and
talent, as when today's celebrity actors and musicians tout their
novels, poetry, or paintings.[3] However, Franklin's case is quite

different, for it is impossible to pin him down to a single area of distinction. He is the poster child for all-around genius, the last true renaissance man: jack of all trades, and master of many. It is hard to believe that so gifted a man as this would find his abilities lacking in any respect.

Nevertheless this is what the experts would have us believe. The editors of *The Papers of Benjamin Franklin* observe that Franklin "was not the mathematician that his friend was," comparing him with the philosopher and clergyman Richard Price, who (like Franklin) speculated on population statistics.[4] A scholar of another eighteenth-century American scientist, Cadwallader Colden, avers that "Franklin could not always follow Colden's reasoning especially in mathematics. . . ."[5] One recent biographer refers to "math, a scholastic deficit he never truly remedied."[6] We find that he "was not sufficiently furnished with a knowledge of mathematics," according to an earlier editor of his papers.[7] Similarly, a Franklin Medal winner described him—in an acceptance speech at the Franklin Institute, no less—as "a *polymath* [a person of greatly varied learning] who excelled at everything *except* mathematics."[8]

If there was an Enlightenment superman, this was Benjamin Franklin: printer, scientist, inventor, author, philosopher, diplomat, and more. As any survivor of the American primary school curriculum can tell you, here was the conqueror of all areas of human achievement. Through hard work and no small share of ingenuity, he managed to overcome a lack of formal education and define the American Dream. And yet, to hear the experts tell it, there remained a gap in Franklin's self-training. The allegation is easy to accept at face value, even comforting. Who among us has never encountered an impediment, an occasional difficulty or even outright failure, in math class? We need our heroes to have flaws, and this one seems plausible enough.

Surely there were gaps in his knowledge, no matter how all-encompassing that polymathic genius may have seemed, yet it is the central thesis of this book that Ben Franklin possessed a mathematical mind. His numerical creations were few, but those that survive

Fig. 1.1. Benjamin Franklin, engraving by A. H. Ritchie (after Charles Nicholas Cochin), no date. American Philosophical Society Library.

demonstrate a feel for number patterns that is unmatched even among many who dedicate their professional lives to mathematics. How much more wonderful, then, that someone who could have devoted only a small portion of his life to the subject would achieve so much in that same pursuit.

A legion of Franklin biographers has misrepresented or misunderstood his fantastic work with magic squares, when not simply ignoring it outright. An exception was Carl Van Doren, whose Pulitzer Prize–winning 1938 biography devoted a few pages to the subject, most of it in Franklin's own words.[9] For his trouble, Van Doren was skewered in a review in *Isis*, the journal of the history of science. The unkind reviewer, I. Bernard Cohen, would go on to become the preeminent science historian of the twentieth century; his articles and books were largely responsible for resuscitating Franklin's scientific reputation in America. The review dismisses Van Doren's biography as "hopelessly inadequate" and claims that the magic squares are given too much attention. Not only are they "of no importance in the development of mathematics," but moreover they represent "no indication of mathematical ability on Franklin's part."[10]

Yet even that distinguished critic would undergo a change of heart. Cohen's own book *Benjamin Franklin's Science* devotes a long passage to the same topic, even going so far as to include a lengthy quote from the same source as Van Doren.[11] This time he sees fit to admit the mathematical importance of magic squares: we must not focus on "obviously practical" goals alone. Magic squares "provide a means of perfecting one's skill in arithmetic." Franklin saw them as "a kind of game or puzzle," which is significant because, as Cohen explains: "The pursuit of mathematics is in any case, according to the German mathematician David Hilbert, like playing a game in which one sets up the rules or operations and sees what results arise from the proper manipulation of the meaningless entities represented by the symbols."[12]

Our object is not to show that Franklin would have identified himself as a mathematician, only that he was adept at the systematic and creative ways of thinking about numbers, arrangements, and relationships that characterize mathematical thought. He was skilled in logical argument, taught himself mathematics as a teenager, and even learned some of the art of navigation on his own. He was a zealous advocate for widespread education in basic accounting skills, repeatedly extolling the virtues of such training for both men and women. His reputation as a universal-genius-*sans*-mathematics is undeserved, as if such a creature were not already an impossibility.

His inner mathematician manifested itself in varied ways. The printing trade, his primary vocation, has mathematical aspects (as we will see in chapter 8). He developed a systematic decision-making technique related to modern *utility theory*, where difficult situations are resolved by means of an algebra for everyday living. For twenty-five years he produced an almanac, a wildly popular pamphlet in a genre that was more typically authored by astronomers and mathematicians. He conceived the most devious magic squares, odd little amusements that must have required considerable facility with number relationships, and these experiments occupied his thoughts periodically for more than half of his long life, as the present book will prove for the first time.

Those magic squares indicate a skill in solving basic algebraic equations, as well as a general comfort with abstract symbols. The latter trait is apparent in other ways, too, such as his use of coded messages and his alphabetic recreations. During the Revolutionary War, Franklin employed simple numerical codes for sensitive communications, though these reveal little of the mathematical sophistication that has come to characterize encryption in more recent times. He attempted to reform the English alphabet, and he corresponded with Noah Webster and Erasmus Darwin on the topic. Several letters from Franklin to his landlady's daughter, and her replies, are even composed in a particular alphabet of his own invention, so it appears that Franklin had no difficulty thinking in abstract, symbolic terms. For what it's worth, his linguistic talents were considerable; he learned languages easily—German, Latin, French, Spanish, and Italian—though he found reading easier than speaking.

It is often said that mathematical and musical proficiency are closely allied; Franklin mechanized the "musical glasses" in his invention of the glass armonica, for which both Mozart and Beethoven composed, and he performed on this instrument. Its very design required knowledge of the relationships between music, geometry, and physics. He created successful lotteries. To describe electrical charge, he appropriated the arithmetic terms *positive* and *negative*, still used for that purpose today. Some say that even the Declaration of Independence bears the mark of Franklin's mathematical side. Thomas Jefferson's original draft asserts, "We hold these truths to be sacred & undeniable, that all men are created equal," and so on. But after incorporating changes from Franklin and John Adams, "sacred & undeniable" was replaced by "self-evident."[13] Like the axioms of Newton or Euclid, each truth is so obvious as to be unprovable, beyond the reach of logical argument. (Among the books Franklin bequeathed to his grandson Ben was a French translation of Euclid's *Elements*, after two millennia the most successful textbook of all time.[14]) It may be no coincidence that the first four of Euclid's five "common" notions also concern equality, such as "Things which equal the same thing also equal one another," though the objects in this case are magnitudes (lengths, areas, or volumes) and not human beings.

While he tended to keep the arguments simple and common-sensical, Franklin had a knack for applying mathematics to areas of scientific and philosophic inquiry where such machinery was as yet rarely used or else completely unknown. His *Observations Concerning the Increase of Mankind and the Peopling of Countries*, an essay composed in 1751 and published four years later, was a landmark in the nascent field of demography, the study of human population statistics. Based on a multitude of factors (such as the heartbreakingly realistic assumption that around half of the children born would not survive to adulthood), he predicted that the population of the colonies would "at least be doubled every twenty years."[15] After some further analysis he allows for the more conservative estimate that it may take twenty-five years. His prognostications were remarkably accurate, especially when one considers that they were made in a time of great social upheaval, and that they belonged to a science that didn't properly exist yet; based on census data from 1790 to 1850, it appears that every twenty years the population increased by 80%, while a complete doubling occurs approximately every twenty-three years, which falls neatly between his two estimates.[16] Franklin's prediction that the population of the colonies would soon outstrip that of England was also borne out, though by then they were colonies no more.[17]

His appears to be a largely intuitive argument, as Franklin refers to the existence of supporting data without actually citing specific quantitative information. Yet careful readers of his almanacs may recognize that, only a year or two earlier, Franklin's *Poor Richard* included population data from three colonies and one European city (broken down in some instances by age, race, and county of residence), and that mortality and doubling-time questions were addressed by him there.[18] Seemingly out of place in a popular almanac, Franklin's ramblings on such topics illuminate some of the mathematical underpinnings of his little excursion into population statistics. As with the magic squares, his mathematical rigor is hidden, but no less real.

That Franklin qualifies as a founder of modern demography can be seen by his influence on Richard Price and Thomas Malthus. Price's analysis of population growth took the form of a personal letter to Franklin, before it appeared in the *Philosophical Transactions of the Royal Society* for 1769. Meanwhile Malthus specifically cites Franklin by name, and his work is acknowledged, in later editions of *An Essay on the Principle of Population*, one of the most important works of social science in all of human history. The Malthusian notion that population may increase exponentially had been hinted at in Poor Richard's almanac, and stated outright in Franklin's *Observations*.[19]

The claim that the number of inhabitants in the colonies would "in another century be more than the people of England" was initially presented, in 1751, in the context of border disputes with the French:

> How important an affair then to Britain is the present treaty for settling the bounds between her colonies and the French, and how careful should she be to secure room enough, since on the room depends so much the increase of her people.

These clashes would soon erupt into the French and Indian War, also called the Seven Years' War, in which both Franklin and a young Colonel Washington served. That same prediction appeared later on in a very different context. An anonymous letter co-written by

Franklin to the London *Public Advertiser* in 1770 used the idea to argue against taxation without representation:

> The British subjects on the west side of the Atlantic see no reason why they must not have the power of giving away their own money, while those on the eastern side claim that privilege. They imagine, it would sound very unmelodious in the ear of an Englishman, to tell him that by the rapidity of population in our colonies, the time will quickly come when the majority of the subjects will be in America; and that in those days there will be no House of Commons in England, but that Britain will be taxed by an American Parliament. . . .[20]

Applying basic mathematics to situations where most of us would not think to do so, he likewise addressed the twin evils of war and slavery. Franklin, a businessman who knew the value of a careful balance sheet, argued in economic terms, circumventing his compatriots' moral ambivalence. Whereas one's views on either issue might be held with a religious zeal, impervious to debate—as in the archaic view that slavery somehow benefited its captives, or in the still popular view that war often serves a greater good—advocates of either position might yield before a purely mathematical argument. To Benjamin Vaughan, the economist and diplomat, Franklin once wrote:

> When will princes learn arithmetic enough to calculate, if they want pieces of one another's territory, how much cheaper it would be to buy them, than to make war for them, even though they were to give a hundred years' purchase? But if glory cannot be valued, and therefore the wars for it cannot be subject to arithmetical calculation so as to show their advantage or disadvantage, at least wars for trade, which have gain for their object, may be proper subjects for such computation; and a trading nation, as well as a single trader, ought to calculate the probabilities of profit and loss, before engaging in any considerable adventure. This however nations seldom do, and we have frequent instances of their spending more money in wars for acquiring or securing branches of commerce, than a hundred years' profit or the full employment of them can compensate.[21]

In a letter to his sister Jane Mecom, he pursues the same line of reasoning. Franklin, who had secured foreign loans to support the Revolution and had extensive personal knowledge of its financial aspects, easily enumerates the specific costs associated with war, adding: "you have all the additional knavish charges of the numerous tribe of contractors to defray, with those of every other dealer who furnishes the articles wanted for your army, and takes advantage of that want to demand exorbitant prices."[22] War simply does not stand up to cost-benefit analysis, according to this philosopher-accountant.[23]

Franklin also argued against slavery using quantitative reasoning. According to his essay on population,

> It is an ill-grounded opinion that, by the labor of slaves, America may possibly vie in cheapness of manufactures with Britain. The labor of slaves can never be so cheap here as the labor of working men is in Britain. Anyone can compute it. Interest of money is in the colonies from 6 to 10 per cent. Slaves, one with another, cost £30 per head. Reckon then the interest of the purchase of the first slave, the insurance or risk on his life, his clothing and diet, expenses in his sickness. . . .[24]

He also sought to turn public opinion based on the sheer size of the slave trade, which was not fully appreciated at that time. In a letter to the *London Chronicle* (1772), he writes that "there are now eight hundred and fifty thousand negroes in the English islands and colonies. . . . [The] yearly importation is about one hundred thousand, of which one third perish" in transit or the "seasoning." He argues by the numbers.[25]

Elsewhere his economic argument is more muted: "Our slaves, Sir, cost us money, and we buy them to make money by their labour. If they are sick, they are not only unprofitable, but expensive."[26] In his later years, Franklin made the transition from small-time slaveholder to outspoken abolitionist, and as president of the Pennsylvania Abolition Society he lobbied Congress on that issue.[27] It would be the last great public act for this former almanac writer who had once intoned: "Nor let me Africa's sable Children see, vended for Slaves though formed by Nature free."[28]

The tendency to think in a precise, rational way about seemingly nonmathematical issues did not fade with age. In his twilight years, Franklin made a rather convincing quantitative argument that the positive qualities of one person do not necessarily translate into similar attributes on the part of their descendants.

In the 1780s, the prospect of establishing a new nobility loomed. American army officers had formed the Society of the Cincinnati, an elite fraternal organization in which membership would automatically pass from father to son. In an era of newly won egalitarianism, such an act was bound to be unpopular. After initial public outcry, membership was to be extended to all who served, not to officers alone. Yet the specter of a hereditary peerage arising so soon after the triumph of democracy over monarchy continued to raise the hackles of a sensitive public and was the subject of much controversy.

Franklin approached the question as an arithmetic problem. Did the sons and grandsons of distinguished veterans deserve to reap the fruits of their fathers' victories? Certainly not, said Franklin, for "descending honours" was a ludicrous notion. While great achievement by an individual may indeed reflect well upon his ancestors, conversely his son shares in only half the honor—as a child is the product of two different families.[29] (The longstanding theory that progeny arose from the seed of one parent alone was by now in its death throes.[30]) Grandchildren share in one-quarter, and so on, until after only nine generations (up to three centuries, he reckons) each descendant will share in "but a 512th part" of that honor. Thus the notion of a hereditary order is not only contrary to the ideals for which the Revolution was fought, it is also contrary to mathematics. (Showing an uncharacteristic absence of tact, Franklin—who amassed several lifetimes' worth of high honors—first introduces this "mathematical demonstration" in a letter to his own daughter.[31]) He opines:

> that all *descending* Honours are wrong and absurd; that the Honour of virtuous Actions appertains only to him that performs them, and is in its nature incommunicable. If it were communicable by

Descent, it must also be divisible among the Descendants; and the more ancient the Family, the less would be found in any one Branch of it. . . .[32]

He refers here to the fact that one-half of one-half of one-half, and so on, moves ever closer to zero. A more nuanced approach to the question of inherited characteristics would have to wait for Charles Darwin (grandson of Franklin's friend Erasmus), Gregor Mendel, and their scientific descendants. Heritable traits are transmitted in a far more subtle and complex way than Franklin suggests; but the point of this example is not that he foresaw any major revolution in genetics, but rather that he felt a "mathematical demonstration" was the appropriate tactic in what was essentially a social debate.[33]

Another simple mathematical idea was used to great effect when Franklin invented the notion of daylight saving time. In a letter to the *Journal de Paris*, he calculates the hypothetical benefit to the city, were his plan to be adopted for roughly half the year.[34] Start with a value of 183 nights. Multiply by seven hours' candle-burning required each night by a household, which accounts for all rooms of the house; then by 100,000, the number of families in Paris. Next multiply this answer by one-half pound, which is the amount of wax and tallow used in an hour. (Lest anyone object to this ad hoc estimate, please note that Franklin grew up in a candle-maker's household!) The final factor is the cost of each pound of these materials, which is around 30 sols. Therefore the cost of all those candles is 1,921,500,000 sols. Since the livre tournois is worth 20 sols, we can divide by 20 to convert the cost to 96,075,000 livres tournois. "An immense sum! that the city of Paris might save every year, only by the economy of using sunshine instead of candles."[35] (One supposes that, were such an idea first proposed today, its implementation would be prevented out of concern for the wax industry.) There's something absolutely poetic in hearing an appeal from spendthrift Poor Richard's alter ego, urging us to save money—a *sol* instead of a *penny* saved—and tricking us into rising early, in the bargain.

The essential idea here is the *multiplication principle*, also known as the *product principle*: if there are 183 days and nights in which the new scheme is to be used, and seven hours of candle-burning to be saved each night, then this amounts to $183 \times 7 = 1{,}281$ hours for each family. If we combine the benefits for all 100,000 families, then $183 \times 7 \times 100{,}000 = 128{,}100{,}000$ hours are at issue, and the calculation continues in this way. Analogous illustrations were employed for entirely different purposes in the pages of *Poor Richard*.[36]

Franklin's proposal is framed as a discovery, not an invention; while anyone who consults an almanac can verify that the sun rises "still earlier every day till towards the end of June," they seem unaware "that he gives light as soon as he rises." Though his suggestion was made in a less than serious manner, this letter to the *Journal* marks the origin of the daylight-saving schemes used today in most of the United States and in other parts of the world. Nothing but the simplest arithmetic, put to serious use.

But the most obvious way in which Franklin embraced mathematical thinking was in his love for the matrix known as the "magic square." That numerical puzzle occupied his thoughts periodically from the early 1730s through the late 1770s, that is, for nearly half a century. As a pastime enjoyed for the better part of a lifetime, by one of the greatest minds of that era, it is surely worth our attention. For the uninitiated, here is a brief introduction to the magic square.

First, a *matrix* (plural *matrices*) is a rectangular array of numbers, letters, words, or other objects. This could be a bookkeeping record, a chart of the tides, or any other arrangement of items, especially abstract symbols or data, into rows and columns. Whenever I teach a course in matrix theory, I wait for the inevitable question: Isn't the definition redundant? Isn't every "array" automatically rectangular? But one can certainly envision arrangements into other shapes. As you'll see in chapter 7, for instance, Franklin constructed a rather ingenious *circular* array. A more familiar example is the infinite *triangular* array called *Pascal's triangle*, named for the French mathematician and religious philosopher

Blaise Pascal (though he was not the first to discover it).[37] The first few rows are

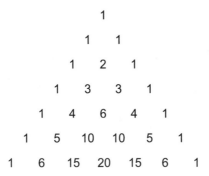

Each entry is equal to the sum of the two entries immediately above it to the left and right. For example, $6 = 1 + 5$ and $15 = 5 + 10$. You could use this rule to work out as many rows as you like, so it really is an infinite triangular array.

In mathematics, though, the term *matrix* does refer specifically to a rectangular arrangement of objects that can be thought of as lying in a grid of smaller squares or rectangles. Matrices defined many aspects of Franklin's life: from the technical aspects of his trade; to the chessboard that he loved; to the weekly record where he kept track of each transgression committed against virtue, in his personal quest for moral perfection (figure 1.2); to the tables of calendars, currency, and kings that filled Poor Richard's almanacs between 1733 and 1758; to his magical squares. The word "matrix" had not yet acquired its modern meaning at that time—in the printing trade, for instance, a matrix (or *matrice*) referred to the mold in which a letter of type is cast—yet it is clear that the concept itself was a motif in Franklin's life. As one commentator puts it in another context, "it is not too much to say that he saw the world through the grid of a case of type."[38]

A *magic square* is a type of square matrix. That means it is the same number of units wide as it is long. We write in the spaces as you would fill in the cells of a crossword puzzle, except that this crossword has no blacked-out cells. In the lingo of puzzlers, it is

TEMPERANCE.							
Eat not to Dulness.							
Drink not to Elevation.							
	S	M	T	W	T	F	S
Temperance	T							
Silence	S	*	*		*		*	
Order	O	**	*	*		*	*	*
Resolution	R			*			*	
Frugality	F		*			*		
Industry	I			*				
Sincerity	S							
Justice	J							
Moderation	M							
Cleanliness	Cl.							
Tranquillity	T							
Chastity	Ch.							
Humility	H							

Fig. 1.2. Sample report card from the *Autobiography*.

more appropriate to refer to it as a cross-*number* puzzle, as we will usually fill the spaces with numerals instead of letters. The goal is to write them so that each line of numbers across, down, or diagonally always totals the same value. For example, in a 5×5 grid there are five rows across, five columns down, and two diagonals (joining opposite corners and passing through the center), or twelve patterns to satisfy in all.

It's easiest to begin with a 3×3 array, like a blank tic-tac-toe board (see figure 1.3). Now you could take the easy way, and just write the number 1 in every space, but most people wouldn't find that to be a very impressive solution. It makes more sense to fill these nine spaces with the first nine counting numbers 1,2,3,4,5,6,7,8,9 in some order. Feel free to put the book down for a few minutes and experiment before reading on.

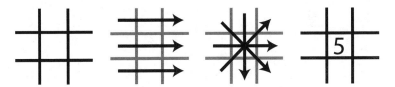

Fig. 1.3. Deriving a 3 × 3 magic square.

Since every row, column, and diagonal must have the same total, it would be helpful to know in advance just what that total should be. The value is readily determined without even knowing which number goes where, as follows. If you were to add up all nine numbers in this little 3 × 3 matrix, the sum would be (in some order) $1 + 2 + 3 + 4 + 5 + 6 + 7 + 8 + 9 = 45$. Therefore, the three equal rows, taken together, add up to 45. That means each row alone sums to 15, so we now know what the "magic sum" will be. (If the puzzle has stumped you till now, try it again with the aid of this clue.)

Once you know that the rows, columns, and diagonals each add up to 15, it's possible to determine what number is placed in the middle of the grid. The key is to look at the middle row across, the middle column down, and both diagonals, all at once. That corresponds to four copies of the "magic sum," so the total of these 12 numbers is equal to four 15's, or 60. But the middle value was included multiple times (four times, to be exact), whilst the other values in our 3 × 3 grid were each included just once. That explains why we got a larger total than 45 this time. Overcounting the middle value three times increased the whole total by 15, so the middle value is equal to 5. There! We finally have one particular entry in place. (That's your last hint before we finish the puzzle!)

Now you need to fill in the middle row, middle column, and both diagonals. Each of these configurations should add up to 15, but each already contains a value of 5 in the middle. To fill out the remaining eight spaces, use pairs: 1 and 9, then 2 and 8, then 3 and 7, and finally 4 and 6. (For extra credit: Why can't the number 1 appear in a corner cell?) There are eight different answers, all equally correct, as shown in figure 1.4.

4	9	2		8	3	4		6	1	8		2	7	6
3	5	7		1	5	9		7	5	3		9	5	1
8	1	6		6	7	2		2	9	4		4	3	8

8	1	6		6	7	2		2	9	4		4	3	8
3	5	7		1	5	9		7	5	3		9	5	1
4	9	2		8	3	4		6	1	8		2	7	6

Fig. 1.4. Eight solutions.

In fact there is really just one answer, in a sense, because all the other solutions are obtained either by rotating the first solution or else by flipping it over (that is, by a mirror reflection). You can check that all three rows, all three columns, and both diagonals add up to the same total. It's magic!

Clearly some creative skill with arithmetic is required by anyone who deals in such puzzles. For centuries, mystics and mathematicians struggled to create ever more impressive magic squares. By the era of pre-Revolutionary America, it was time for the master of the magic square to unveil his work.

Notes

1. Letter from Franklin to his sister Jane Mecom, Sept. 20, 1787. Albert Henry Smyth, ed., *The Writings of Benjamin Franklin, Collected and Edited with a Life and Introduction by Albert Henry Smyth*, New York: The Macmillan Company, 1905–7, Vol. 9, p. 613.

2. Larry W. Phillips, ed., *Ernest Hemingway on Writing*, New York: Touchstone, 1999, esp. pp. 48–49. Hemingway also states that "having books published . . . is even worse than making love too much" (p. 55).

3. Hopefully this comment does not apply to a mathematician who writes history.

4. Leonard W. Labaree *et al.*, eds., *The Papers of Benjamin Franklin*, New Haven and London: Yale University Press, 1972, Vol. 16, p. 81. Henceforth this source will be abbreviated *Papers*.

5. Unsigned preface to *The Letters and Papers of Cadwallader Colden,* New York: Printed for the New York Historical Society, 1918–37, Vol. 1 (1711–1729), p. vi. The assertion might be correct. However, it may be that Franklin claimed ignorance simply to avoid insulting Colden, who had asked for feedback on a mathematical manuscript that suffered from many shortcomings (according to James Logan, whose qualifications are not in dispute).

6. Walter Isaacson, *Benjamin Franklin: An American Life*, New York: Simon & Schuster, 2003.

7. Smyth, *The Writings of Benjamin Franklin,* Vol. 1, p. 73.

8. The speaker was computer scientist and mathematician Donald Knuth, a man of Franklin-level brilliance who won the award in 1988. To his everlasting credit, Knuth has assured me that this assertion will be recanted in future printings of his *Selected Papers on Computer Science*, after having read one of my articles on Franklin.

9. Carl Van Doren, *Benjamin Franklin*, New York: The Viking Press, 1938, pp. 143–146.

10. I. Bernard Cohen, review, *Isis*, Vol. 31, No. 1, 1939, pp. 91–94. Later, Cohen became editor of the journal.

11. Like Van Doren, Cohen allocates approximately one thousand words to the topic.

12. I. Bernard Cohen, *Benjamin Franklin's Science*, Cambridge: Harvard University Press, 1990. The chapter I have quoted was originally published as "How Practical Was Benjamin Franklin's Science?" in *The Pennsylvania Magazine of History and Biography*, Oct. 1945, pp. 284–293. It is worth noting that in 1947, Cohen reviewed another work by Van Doren, the collection *Benjamin Franklin's Autobiographical Writings*, which also included material on the magic squares, largely identical to that in his earlier book; but this review made no mention of the fact, and even praised a "model of scholarly workmanship." *Isis*, Vol. 37, No. 1–2, 1947, pp. 85–86.

13. Carl Becker (*The Declaration of Independence: A Study in the History of Political Ideas*, New York: Vintage Books, 1970) and Walter Isaacson (*Benjamin Franklin: An American Life*, 2003) identify Franklin as the likely author of that essential emendation. Others believe Jefferson himself to be responsible for the change. However, the latter argument is based on the belief that this particular alteration was made in Jefferson's handwriting—which does not change the fact that Jefferson incorporated suggestions of Adams and Franklin, which he may have then written out himself.

14. "List of Books for B. F. Bache," manuscript in the Library Company of Philadelphia. The influence of Newtonian science on Franklin and Jefferson is described in I. Bernard Cohen, *Science and the Founding Fathers*, New York: W. W. Norton and Company, 1997.

15. It is interesting to note that, of Franklin's three children, two made it to adulthood. Whether this personal experience influenced his 50% statistic is unknown, for he does not explain it, other than to say that he assumes eight children to a marriage, of whom four survive.

16. The census figures themselves show an increase varying between 77.04% and 84.28%, but a standard statistical approach leads one to the conclusion that 80.45% is the correct figure. (For the technically minded, I have used an exponential regression model fit to eight data points, with $R^2 > 0.998$.)

17. In the 1830s, the American population did surpass the British.

18. *Poor Richard Improved ... for the Year of Our Lord 1750*, Franklin and Hall, 1749.

19. That he believed growth to be exponential is clear from the fact that he refers to a constant doubling time: "doubled every twenty years," "doubling ... once in twenty-five years." Contrast this with *arithmetic* growth, where the same number of inhabitants is added every year, so that it takes longer and longer for the population to double as time passes. (Perhaps Franklin's awareness of exponential growth originated in compound interest calculations.)

20. *Papers*, 1973, Vol. 17, pp. 5–8. There is some controversy over the degree of collaboration between these coauthors: see Carla H. Hay, "Benjamin Franklin, James Burgh, and the Authorship of 'The Colonist's Advocate' Letters," *The William and Mary Quarterly*, 3rd Series, Vol. 32, No. 1, 1975, pp. 111–124. The inclusion in this letter of a detail from Franklin's earlier *Observations* adds some weight to the claim that he is a coauthor.

21. *The Writings of Benjamin Franklin*, Vol. 9, pp. 676–677.

22. *Ibid*, pp. 612–613.

23. Yet another instance where Franklin makes an economic argument against war is in a letter to an unnamed correspondent, in William Temple Franklin, *Memoirs of the Life and Writings of Benjamin Franklin*, London: Henry Colburn, Vol. 2, 1817, pp. 106–107.

24. *Observations Concerning the Increase of Mankind*, 1751 (published 1755).

25. *The London Chronicle*, June 18–20, 1772, as reprinted in the *Papers*, 1976, Vol. 19, pp. 187–188; also pp. 112–115 for his source, the abolitionist Anthony Benezet, explicitly acknowledged on p. 269.

26. Letter dated January 15, 1766, in *Papers*, 1969, Vol. 13, pp. 44–49. The letter is signed "Homespun," but it was attributed to Franklin by his grandson.

27. Joseph Ellis, *Founding Brothers: The Revolutionary Generation*, New York: Alfred A. Knopf, 2000.

28. *Poor Richard Improved* for 1752, paraphrasing Richard Savage's *Of Public Spirit in Regard to Public Works* (1737).

29. He points out that the commandments instruct one to honor one's father and mother, not necessarily one's children. I am reminded of Plutarch: "It is indeed a desirable thing to be well descended, but the glory belongs to our ancestors."

30. This theory (preformationism) was itself divided into two opposing camps: the spermists and the ovists. For example, the former held that the sperm cell contained a tiny but complete person, and within each tiny male *homunculus* were even smaller *homunculi*, and that the entirety of the future human race was stacked like Russian dolls, ad infinitum. The successor theory recognized that both parents contributed essential ingredients.

31. These ideas are described in detail in a letter to Sarah Bache (Jan. 26, 1784), and referred to briefly in another to George Whatley (May 23, 1785). Smyth, *Writings*, Vol. 9, pp. 161–168 and 331–339. For a similar (and earlier) mathematical application, see *Poor Richard Improved* for 1751.

32. Letter to G. Whatley; see note 31.

33. A similar explanation is sometimes given as an objection to the use of DNA testing in order to identify remote ancestors (see, for example, in *Time*, July 11, 2005). If you learn of a tenth-generation ancestor of great importance, that still leaves a thousand others who made equal contribution to your being.

34. *Writings*, Vol. 9, pp. 183–189.

35. A desire to conserve resources remains the driving force behind the use of daylight saving time (DST). During a major oil crisis in the 1970s, the federal government temporarily mandated its use year round. Ironically, oil was not a new issue here; Franklin's thoughts on the subject were initially inspired, not by candles, but by the question of a new oil lamp then in use: "whether the oil it consumed was not in proportion to the light it afforded, in which case there would be no saving in the use of it." The question is ever more relevant in an age in which we are burning the candle at both ends. (A new law mandates four extra weeks of DST, to begin in 2007.)

36. Poor Richard's almanacs include quite a few examples of such repeated multiplication, as we will see in chapter 3.

37. It is true that, by rotating and stretching the configuration, Pascal's triangle can be expressed as an infinite *rectangular* array:

$$
\begin{array}{cccc}
1 & 1 & 1 & 1 \ \ldots \\
1 & 2 & 3 & 4 \ \ldots \\
1 & 3 & 6 & 10 \ \ldots \\
\vdots & \vdots & \vdots & \vdots
\end{array}
$$

and it is so written in some of the early European works. However, in a much older Chinese incarnation—and in the version most of us learned in school—it is built on an acute angle (less than ninety degrees) opening downward as we have drawn it. Indeed, Pascal's triangle is usually abbreviated as a *finite* array in classroom settings, and this finite array is always triangular (never rectangular).

38. Christopher Looby, "Phonetics and Politics: Franklin's Alphabet as a Political Design," *Eighteenth-Century Studies*, Vol. 18, No. 1, 1984, pp. 1–34.

2 A Brief History of Magic

The turtle, pleasingly to float
asleep upon the sea;
But when it's catch'd by men in th' boat,
it wakes immediately.
  —The fables of young Æsop (1700)

*A*ccording to generally accepted myth, it all began with a turtle. Not the legendary world-bearing beast, that Atlas of reptiles who has carried the earth on his back since time began, but an unassuming little creature void of extraordinary qualities save for the mystical markings on his spotted lower shell.

Around 2200 B.C.E., so the story goes, such a beast crawled out of the Lo River and fatefully crossed the path of Emperor Yu. Other sources put the discovery closer to 2800 B.C.E. and attribute the design to Fu Xi, a legendary founder of Chinese civilization.[1] Adorned with the images of counting beads or knotted ropes—think of the way numbers are depicted on dice and dominoes, or of the counters on an abacus—this special turtle inspired the *lo shu*, the 3×3 magic square named for the river from which it ascended.[2] (See figure 2.1.) Notice that white beads are used to represent odd numbers, black beads to show the evens. Convert these symbols to modern notation and you will see that this matrix is identical to our tic-tac-toe solution from the end of chapter 1.

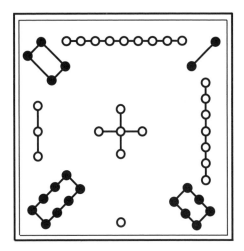

Fig 2.1. The *lo shu* 3 × 3 magic square.

Uncritical commentators have accepted these early dates of discovery without question, but some healthy skepticism is in order; the fact is, we only know of this tale through much later sources that were composed long after the purported event. Suffice to say that it is possible to accept the existence of Yu and even the existence of turtles without giving too much credence to other details of the story. I am inclined to believe that this legend was inspired by the ancient practice of *scapulimancy*, the art of reading patterns on animal bones or shells for the purpose of divination (that is, prophecy and other insight by magical means). Often an animal's shoulder blade was used, hence the derivation from the root word *scapula*. The bones or shells were usually leftovers from dinner, but the uncooked were also subject to this form of fortune telling. Some shells bear markings that were intentionally placed. Quite recently, archaeologists unearthed what may be the earliest known evidence of a human writing system—symbols carved on tortoise shells more than eight thousand years ago, possibly the remains of musical instruments used in religious ritual.[3] As it happens, these artifacts and similar, more recent ones were found in China, home of our mathematical turtle.

Knowledge of the lo shu matrix and belief in its supernatural properties would spread over a vast portion of the planet. By the end of the first millennium C.E., Islamic mathematicians had investigated such "harmonious distributions of numbers in squares" of various sizes.[4] In 1275, Chinese author Yang Hui published many examples which he attributed to far earlier texts.[5] Tibet, India, Japan, Thailand, Malaysia, and—by way of Islam—Europe and Africa, all would inherit the magic square tradition. It is difficult to imagine another concept of such little apparent use that spread so widely, to so many cultures, in so many different forms.[6]

An infectious, evolving idea of this sort is called a *meme*, the mental analog of a gene. A meme can be a catchy tune, a gesture, or any other unit of cultural information that passes from one brain to another. It is to culture what the gene is to biology. Modeled after genetics, memetics is a relatively new field of study, and at present it is considered quite controversial. At issue is whether the concept of a meme is meaningful; that is, will memetics predict anything that could not have been learned by other means?[7]

Whether or not science ultimately accepts memetics or dismisses it as just another fad, the magic square is a perfect example of a meme. As a replicator, its reproductive success cannot be denied; in twenty-first century America, one finds magic squares in puzzle magazines, in elementary school classes, and in the New Age section of the bookstore. Unlike the ability to build a fire, or the coining of a word that means "look out, there are tigers nearby," the magic square meme does not survive based on some intrinsic usefulness. It's just catchy, like a cold.

The self-replicating meme mutates and evolves, just like a gene. As the idea of the magic square spread throughout the world, it adopted many shapes and sizes: 4×4 squares, 5×5 squares, and on and on; later still, we see magic rectangles, triangles, hexagons, and more (figure 2.2).[8] The magic square was used in incantations and spells, emblazoned on amulets and talismans and plates, and incorporated into astrology, witchcraft, and other prescientific beliefs inherited from Babylon, Greece, and India. These squares were also of mathematical interest, for according to one expert, "they constituted a major preoccupation for many brilliant scholars in the Middle East and India" and, of course, China.[9]

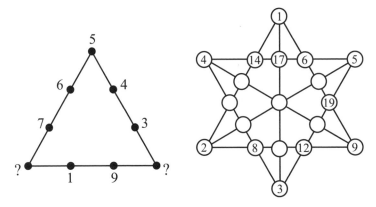

Fig 2.2. Modern descendants: magic triangle and hexagram, or six-pointed star. Fill in the missing values so that each line adds to the same sum.

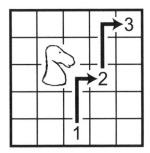

Fig 2.3.

Once we graduate beyond the 3×3 case, there are many different methods for building a magic square. Consider just one early technique, which will amuse anyone who has played a little chess. Start with an empty square grid that has odd dimensions, at least 5×5. Based on the 3×3 magic square, it is reasonable to assume that we might begin by placing the number 1 in the middle of the bottom row. (This is just an inspired guess, but analogy can be a reliable guide in this business.) From that cell, move up two spaces and right one space, like the knight on a chessboard, to find a home for the number 2, as in figure 2.3. So far, so good. In the 3×3 magic square we'd run out of room now, but for 5×5 (or 7×7 or larger),

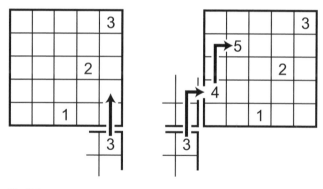

Fig 2.4.

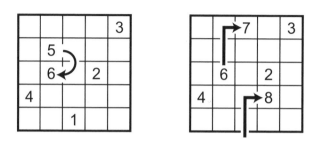

Fig 2.5.

we could repeat the step once more: go up two and over one, to place the number 3 in the appropriate cell.

Alas, in a 5 × 5 array we have now landed in the northeast corner, so another knight's move in the same direction is impossible. But if you imagine that the chessboard continues off the top edge by tele-porting to the bottom, then it still makes sense to move "up" two cells (figure 2.4). Likewise you can jump off the right-hand side and end up on the left. That allows us to continue making knight's moves for a little while longer.

Eventually, however, our brave knight encounters a more serious obstacle: the space we'd like to land in is already occupied by an-other number! When that happens, simply place the next value (which is 6, in our example) immediately below the previous entry, then resume knight-skipping (figure 2.5).

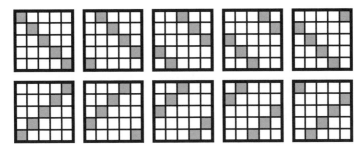

Fig 2.6. Broken diagonals.

You can finish the magic square by hand. (The final answer for our 5 × 5 square can be found in the endnotes.[10]) Just keep moving your knight two units north and one unit east each time. If you come to an edge, skip across to the opposite edge. If a seat is occupied, shift south by one unit instead.

It is easy to check by hand that the rows and columns all add up to 65. The diagonals both work out fine as well: the forward diagonal traces from the lower left corner to the upper right, and the back-slash connects the remaining corners. But wait, there's more! Not only is this a magic square, but we have accidentally created a *pandiagonal* or *diabolic* square.[11] That means even if you draw diagonals that break across the edge of the square, they still add up to the correct value of 65 (figure 2.6).

The knight's move square and similar "uniform step" methods work perfectly well as a solution to the problem of building magic squares of odd dimensions. If the side length is an even number, the construction is a bit tougher; and among the even-numbered sizes, multiples of 4 are easier than 6, 10, 14, and so on. Hundreds of articles have been published on the subject of magic squares and their creation, and that statistic doesn't count the vast efforts by legions of unpublished amateur mathematicians throughout the world.

What I find most fascinating about magic squares, apart from their mathematics and their simple tenacity as a perpetually occurring artifact, is the propensity of their advocates toward religious mysticism.

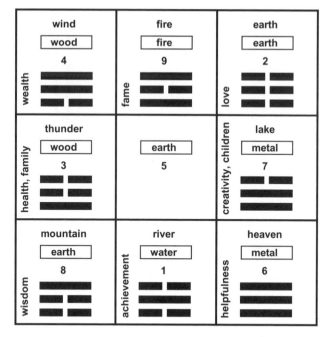

Fig 2.7. Feng shui guide.

The *I Ching* (classic of changes) is one of the foundational texts of Taoism. One of the most ancient living religions, Taoism claims millions of adherents worldwide. It is not known precisely when the lo shu entered the tradition of the *I Ching*, but those who assert the great antiquity of the magic square rely on unverified dates given by later Taoist commentaries. The diagram itself, expressed in trigrams, is also used in the practice of *feng shui* (figure 2.7). Not simply a method for arranging one's surroundings in harmony, feng shui represents an all-encompassing philosophy of life, and its roots go back to the very first magic square.

Such connections are not unique to Chinese religion and philosophy. Prescribed for a wealth of purposes by Persian and Arab authors throughout the Islamic world, and by Hindu mathematicians in India, magic squares were displayed on supernatural charms, engraved on plates, and worn on decorative rings.[12] (This is unsurprising, as these civilizations excelled in mathematics during the period of medieval scientific dormancy suffered in Europe.) These could be written numerically, or else composed in Hebrew letters whose numeric interpretation yields a magic square. Often the scribe who copied the square from one mystic text to the next was ignorant of its mathematical meaning, and a value or two might be carelessly changed, the meme-equivalent of a harmful genetic mutation. Some numerals might be altered on purpose, as camouflage, so as to hide their meaning from the uninitiated or unworthy. Squares of size 3 through 9 corresponded to each of the seven astrological "planets": Saturn, Jupiter, Mars, Sun, Venus, Mercury, and Moon. Each square encodes the name of God in some way. Beliefs in such arcana were incorporated into the so-called "Christian Cabala," a loosely defined system of beliefs that forms the basis for much New Age mysticism in the West (especially astrology, witchcraft, and numerology). One sacred text even attributes the first magic square to Adam, who learned it from God Himself and then passed the knowledge down through select descendants.[13]

An Islamic manuscript specifies the uses of each square. The 3×3 Saturn square, for instance, can be used to guarantee safety for a woman giving birth. Simply write it on cotton at an astrologically auspicious time and apply it to her right hip during labor. Or, etch the same matrix on a lead plate and carry it with you for general protection from harm. If the plate is created at the wrong time, however, the luck is reversed. Similarly detailed instructions are given for the larger magic squares.[14] While it is easy to be amused by what is today considered pseudoscientific speculation, we must recognize the need to impose order on a complex world. After all, the very same motivations drive both science and pseudoscience today.

Some magic square patterns were ascribed to Plato and Archimedes, but these fanciful attributions have no basis in fact.[15] It appears that the ancient Greeks, for all their mathematical virtuosity, were not familiar with this particular aspect of numerology. Thus it was not until relatively late that Europe, through a variety of routes, was finally infected with the magical meme. It is likely that European pilgrims and Crusaders obtained magic plates in the Holy Land. These were festooned with planetary squares, offering protection against the plague and other threats. (And why not? That was certainly the best protection available at the time.[16]) Less speculative sources are certainly known. A treatise by the four-teenth-century Byzantine scholar Manuel Moschopoulos relates Islamic methods, translated from Arabic to Greek.[17] Another source was Muslim Spain; one expert cites a specific "Arabic [text] of Moorish origin."[18] Thus it seems that this information traveled independently from both Asia Minor and northern Africa, penetrating southern Europe from east and west. A separate line of transmission came directly from the Far East when Simon de la Loubère, Louis XIV's envoy to Siam (present-day Thailand), published an account of Siamese society, culture, and technology in the 1680s.[19]

By de la Loubère's time, however, the concept was already long a part of the fabric of western occultism, largely through the work of alchemist and astrologer Henricus Cornelius Agrippa von Nettesheim.[20] He gave each planetary seal in both numbers and letters, together with instructions for its use.[21] These are illustrated in figure 2.8. Followers of Agrippa, citing unlikely Egyptian and Pythagorean forebears, identify the 2×2 magic square—a logical impossibility—with "imperfect matter," and assign "to God the square of only one cell, the side of which is also a unit, which multiplied by itself, undergoes no change."[22] Meanwhile another class of matrices had entered the western occult tradition, and though these are unrelated to the lo shu, they too are frequently referred to as "magic squares." Mystically imbued like the planetary squares but written in a Latin alphabet and lacking any arithmetic significance, they were fairly well known in the lands of the former Roman Empire. The best known of these is the SATOR square, below. The same

S	A	T	O	R
A	R	E	P	O
T	E	N	E	T
O	P	E	R	A
R	O	T	A	S

cryptic message *Sator arepo tenet opera rotas* is expressed by reading the matrix in all four directions: across, down, backward, or upward. It has been alleged that this palindromic cipher was used as a password by which early Christians identified one another, since the letters can be rearranged into a cross that reads *Pater Noster* (Our Father) both across and down, with leftover letters consisting of a,a,o,o (the alpha and the omega, naturally).[23] Alphabetic magic squares written on church walls, and numeric magic squares in Agrippa's Christian Cabala, continue the religious thread that led from China to the Middle East and on to Europe.

Renaissance artist and geometer Albrecht Dürer, likely inspired by the same alchemical sources as Agrippa, used a slightly different incarnation of the Jupiter square in his *Melencolia I* (1514) (see figure 2.9). This 4×4 matrix can be obtained from Agrippa's by flipping over the latter square and then exchanging its second and third columns.[24]

Based on Dürer's other works, one would expect this scene to be decorated with the tag "1514 AD." In his equally famous engraving *The Knight, Death and the Devil,* a typical legend (shown in figure 2.9, far right) lies unobtrusively in one corner. The logo shows D within A for the artist's initials, or perhaps for *anno Domini.* The signature is curiously absent in *Melencolia.* However, it is well known that the bottom row of his magic square serves the same purpose, for the numerals 15,14 appear there. (Dürer's initials comprise the remainder of the bottom row: 1 for A, 4 for D.[25]) This is just the first of many instances in which the magic square has infected art, architecture, music, and literature (see, for example, figure 2.10[26]). To take a few examples from the past century alone: The Swiss painter Paul Klee, who like Dürer is regularly included in lists of the greatest painters of all time, based many of his works on the magic square, replacing numbers with colors. Architectural theorist Claude Bragdon formulated ornamental designs built around the patterns traced by connecting the dots in a magic square, following successively 1,2,3, and so on; he had an especial interest in the squares of Benjamin Franklin. Composer John Cage brought the magic square to music, and several novels have explored the same

**Saturn's square,
made of lead**

4	9	2
3	5	7
8	1	6

ד	ט	ב
ז	ה	ג
ח	א	ו

birth

**Iouis (Jupiter)
silver**

4	14	15	1
9	7	6	12
5	11	10	8
16	2	3	13

ד	יד	יה	א
ט	ז	ו	יב
ה	יא	י	ח
יו	ב	ג	יג

wealth, love

**Solis (Sun)
gold**

6	32	3	34	35	1
7	11	27	28	8	30
19	14	16	15	23	24
18	20	22	21	17	13
25	29	10	9	26	12
36	5	33	4	2	31

ו	לב	ג	לד	לה	א
ז	יא	כז	כח	ח	ל
יט	יד	יו	יה	כג	כד
יח	כ	כב	כא	יז	יג
כה	כט	י	ט	כו	יב
לו	ה	לג	ד	ב	לא

fame, power

**Mercurij (Mercury)
silver, tin, or brass**

8	58	59	5	4	62	63	1
49	15	14	52	53	11	10	56
41	23	22	44	45	19	18	48
32	34	35	29	28	38	39	25
40	26	27	37	36	30	31	33
17	47	46	20	21	43	42	24
9	55	54	12	13	51	50	16
64	2	3	61	60	6	7	57

ח	סב	סג	ה	ד	נט	נח	א
מט	יה	יד	נב	נג	יא	י	נו
מא	כג	כב	מד	מה	יט	יח	מח
לב	לד	לה	כט	כח	לח	לט	כה
מ	כו	כז	לז	לו	ל	לא	לג
יז	מז	מו	כ	כא	מג	מב	כד
ט	נה	נד	יב	יג	נא	נ	יו
סד	ב	ג	סא	ס	ו	ז	נז

wealth, knowledge

Fig 2.8. Agrippa's planetary magic squares, in Hindu-Arabic numerals and Hebrew letters. Some authors reverse the order of the planets, and one of the squares is sometimes replaced by another of the same size.

Martis (Mars)
iron

11	24	7	20	3
4	12	25	8	16
17	5	13	21	9
10	18	1	14	22
23	6	19	2	15

יא	כד	ז	כ	ג
ד	יב	כה	ח	יו
יז	ה	יג	כא	ט
י	יח	א	יד	כב
כג	ו	יט	ב	יה

war

Veneris (Venus)
silver

22	47	16	41	10	35	4
5	23	48	17	42	11	29
30	6	24	49	18	36	12
13	31	7	25	43	19	37
38	14	32	1	26	44	20
21	39	8	33	2	27	45
46	15	40	9	34	3	28

ד	לה	י	מא	יו	מז	כב
כט	יא	מב	יז	מח	כג	ה
יב	לו	יח	מט	כד	ו	ל
לז	יט	מג	כה	ז	לא	יג
כ	מד	כו	א	לב	יד	לח
מה	כז	ב	לג	ח	לט	כא
כח	ג	לד	ט	מ	יה	מו

fertility

Lunae (Moon)
silver

37	78	29	70	21	62	13	54	5
6	38	79	30	71	22	63	14	46
47	7	39	80	31	72	23	55	15
16	48	8	40	81	32	64	24	56
57	17	49	9	41	73	33	65	25
26	58	18	50	1	42	74	34	66
67	27	59	10	51	2	43	75	35
36	68	19	60	11	52	3	44	76
77	28	69	20	61	12	53	4	45

ה	נד	יג	סב	כא	ע	כט	עח	לז
מו	יד	סג	כב	עא	ל	עט	לח	ו
יה	נה	כג	עב	לא	פ	לט	ז	מז
נו	כד	סד	לב	פא	מ	ח	מח	יו
כה	סה	לג	עג	מא	ט	מט	יז	נז
סו	לד	עד	מב	א	נ	יח	נח	כו
לה	עה	מג	ב	נא	י	נט	כז	סז
עו	מד	ג	נב	יא	ס	יט	סח	לו
מה	ד	נג	יב	סא	כ	סט	כח	עז

health, protection against enemies

Fig 2.8. (continued)

Fig 2.9. Above: Albrecht Dürer, *Melencolia I,* Rosenwald Collection, image ©
2006 Board of Trustees, National Gallery of Art, Washington. Right: Detail
from Hermann Schubert's *Mathematical Essays and Recreations*, 1898.
Far right: A more typical example of Dürer's signature.

Fig 2.9. (continued)

theme, most recently Steve Martin's *The Pleasure of My Company*.[27] It is safe to say the idea has infiltrated many diverse aspects of modern culture.

Having invaded the West, magic squares continued to attract the attention of not only mystics and pseudoscientific healers, but also serious mathematicians. Among these were Girolamo Cardano and Blaise Pascal, two of the biggest names of sixteenth-century Milan and seventeenth-century France, respectively. A younger contemporary of Agrippa and Dürer, Cardano was the son of a lawyer and university mathematics lecturer who advised Leonardo da Vinci. You may recall the quadratic formula, which is used to solve equations in which a variable is squared; Cardano published a comparable solution for cubic and quartic equations, though these were both the work of other mathematicians. He was also a pioneering probabilist and, not coincidentally, a gambling addict. Likely it was the interplay of mathematics and astrology that attracted him to the magic squares, for he was no stranger to the stars; at one point he was even jailed for heresy, having cast the horoscope of Jesus Christ, among other offenses.[28]

Like Cardano, Blaise Pascal chose not to follow his father into the legal profession. Pascal invented a mechanical calculating machine, helped lay the foundations of calculus, and did research in geometry

Fig 2.10. The Judas kiss and accompanying magic square in Josep Maria
Subirachs's Passion façade (1988), part of Antoni Gaudí's *Temple de la
Sagrada Família* (Temple of the Holy Family) in Barcelona. Photos by Gloria
Fluxà Thienemann. In a typical 4 × 4 magic square, the sum of a row, column,
or diagonal is 34, but here the numbers have been altered to make that sum
equal 33, symbolizing the life of Christ. This magic square bears a striking
similarity to Dürer's, if the latter matrix is rotated 180 degrees (lower right).
Only four entries differ, each of these by 1.

and other areas of mathematics, much of this before his teenage years were exhausted. He, too, worked in the nascent science of probability, where the numbers in his arithmetical triangle play no small role, and he is acknowledged as one of the true founders of modern probability. A simple example illustrates the connection between the numerical triangle and games of chance. Suppose a fair coin is flipped three times. You may obtain three heads (denoted HHH), or two heads and one tails (this can happen three ways: HHT, HTH, THH), or one heads and two tails (three ways: HTT, THT, TTH), or three tails (one way: TTT). These events occur in 1,3,3,1 ways, respectively. This corresponds to a row of Pascal's triangle that reads 1 3 3 1. (See chapter 1.) In all there are $1 + 3 + 3 + 1 = 8$ ways the experiment can occur. Therefore if you want to know the probability that you get three heads, it is $1/8 = 12.5\%$; of getting exactly two heads, $3/8 = 37.5\%$; and so on. (If the coin is flipped four times, use the next row of the triangle instead.)

Pascal was also an accomplished physicist, philosopher, and theologian. Most impressive of all is that his multifarious accomplishments filled such a brief lifetime; when Pascal died in 1662, he was only thirty-nine years old. His mathematical discoveries are wide ranging, but less well known is the fact that he dabbled in magic squares as well.[29]

Further mutations of the magical meme continued unabated. The three-dimensional magic *cube* was born. Bernard Frénicle de Bessy's landmark study of the magic square was published in 1693, the same year that de la Loubère's Siamese travelogue was translated into English. In 1704 a Brussels clergyman hit on the idea of building magic squares using repeated entries, inventing an early version of what has come to be known as the Latin square.[30] Chronologically, this finally brings us to the advent of the greatest American genius of the eighteenth century.

The year was 1706 by our reckoning, though it is more correct to write 1705/6, for England and its colonies had not yet adopted the new calendar. Ben Franklin was the fifteenth of Josiah Franklin's seventeen children, the tenth by second wife Abiah Folger. The family lived in the Puritan Boston of the Mathers. A candle maker and soap boiler by trade, Josiah hoped his youngest son would enter the clergy. Ben was sent to grammar school at the age of eight but stayed less than a year; it seems that the high cost of a scholarly education, coupled with the prospect of little financial return, caused Josiah to question the wisdom of his plans. Next the boy attended a school run by George Brownell, well known for his success in teaching writing and arithmetic.[31] Alas, Franklin recounts in his autobiography: "... I acquired fair writing pretty soon, but I failed in the arithmetic, and made no progress in it." Thus ended his formal education.

Now ten years old, Ben began working with his father, but it soon became clear that he was not born for the candle trade. His longing for a life at sea alarmed Josiah, who began to look in earnest for a more acceptable calling. After a brief trial period working for his cousin, a cutler, the twelve-year-old boy instead became an apprentice to his half-brother James, a printer, in 1718.[32]

The following year, an American edition of *Hodder's Arithmetick* appeared. It was the first English-language textbook published in the colonies that was devoted solely to mathematics. What appears to have escaped notice is that this book was produced in the very same printing house where Ben Franklin was employed.[33] Yet if the young apprentice was involved in its production, composing type and running the press, it probably made no impression on him other than to highlight an appalling gap in his academic preparation. It was not until three years later that he would rectify this shortcoming. He recalls: "And now it was [at age 16] that being on some occasion made ashamed of my ignorance in figures, which I had twice failed in learning while at school, I took Cocker's book of arithmetic, and went through the whole by myself with great ease."[34]

Textbooks attributed to Edward Cocker were not limited to arithmetic but also covered writing and other topics, though there is some debate regarding their true authorship. Decades after their first appearance, one could still purchase Cocker's *Spelling Books* and *Arithmetick* in Franklin's Philadelphia printing office.[35] Their popular status is further attested by a verse composed in 1713 by

Josiah's brother Benjamin, who urges his namesake nephew to write "till thou excel great Cocker with thy quill."[36] That wish was surely granted.

The mathematics text in particular was enormously successful, continuing through many editions, and even inspiring a popular expression: the phrase *according to Cocker* refers to something that is indisputably correct, in accordance with generally accepted rules, or "by the book."[37] This textbook covers the basic operations of addition, subtraction, multiplication, and division, explaining how best to accomplish each feat; his methods correspond to those found throughout Franklin's papers, scribbled in margins and on the back of old correspondence. Cocker's book covers fractions—also called "broken numbers"—as one would expect in such a primer. These are *vulgar* or common fractions, to be distinguished from decimal fractions. There are no recreational topics here, as the emphasis is resolutely practical: calculating exchange rates, simple interest, and weights and measures. One exercise in this last category reads:

> A grocer bought 2 chests of sugar, the one weighed neat 17 hundredweight, 3 quarters, 14 pounds at 2 pounds 6 shillings 8 pence per hundredweight, the other weighed neat 18 hundredweight, 1 quarter, 21 pounds at $4\frac{1}{2}$ pence per pound, which he mingleth together, now I desire to know how much a hundredweight of this mixture is worth?[38]

To solve this problem, you need to know how the units are related: 28 pounds in one quarter, four quarters in one hundredweight. Thus a British *hundredweight* is really 112 pounds, in case you were worried that any of this would be easy. Also, one pound sterling equals 20 shillings, and each shilling is equal to 12 pence. (If you are traveling to England, rest assured that the system has since been decimalized. Now a pound is worth 100 pence, not 240.) A good working knowledge of unit conversion was no small necessity for any merchant or tradesman who lived in the Empire of pounds, shillings, pence, and farthings, to say nothing of groats and nobles and crowns, and this presumes that one is dealing with English currency alone. Likewise, dozens of terms for dry measure, length, time, and other categories are detailed in *Cocker's Arithmetick*, and even so, it is by no means an exhaustive guide to the bewildering array of weights and measures then in simultaneous use.

The enduring myth of Franklin as an eminently practical philosopher is bolstered by his mention of the *Arithmetick* in his memoir. When Franklin wonders of what use a magic square may be, he is expressing the inner voice of his earliest mathematical training.[39] For Cocker's preface makes plain that the author aims not to be "abstruse or profound" but rather strives to make his guide useful, with the particular needs of merchants a major inspiration. He pointedly disparages "pretended numerists" who "propound unnecessary questions."

But while the *Arithmetick* celebrates the concrete aspects of number, another English book read by the sixteen-year-old Franklin is broader in scope. John Locke's *An Essay Concerning Human Understanding* explores the nature of ideas, perception, and truth. Locke argues that self-evident truths are not limited to statements of mathematics, but may admit a broader class of assertions. He does rely on various numeric examples to explain what is meant by an axiom, or *maxim*, a simple example being "one and one is two." Indeed, knowledge and even thought itself is based in number:

> Amongst all the ideas we have, as there is none suggested to the mind by more ways, so there is none more simple than that of *unity*, or one. . . . [It is] the most universal idea we have: for number applies itself to men, angels, actions, thoughts, every thing that either doth exist, or can be imagined.[40]

Combining unity with itself repeatedly, we can obtain other numbers, restricted only by our ability to put names to those amounts. Locke describes an indigenous people of Brazil whose number-names do not exceed five.[41] Similarly, he claims, American colonists confronted with the challenge of expressing large numbers such as one thousand "would show the hairs of their head, to express a great multitude which they could not number; which inability, I suppose, proceeded from their want of names." They could count to twenty, but not to one thousand. (I find this example to be curiously chauvinistic. Were only émigrés so afflicted?) Locke provides some helpful terminology, defining numbers from *milion* [sic] and *bilion* to *octilion* and *nonilion*, so that one may express values whose decimal expressions run to dozens of digits.[42] He demonstrates how such terms may be used to name a sixty-digit number.

Locke also addresses the notion of infinity: infinite space, infinite time (eternity), and quantities both large and small. Though there exist infinitely *many* numbers, says he, there is no infinite number. He considers the infinitely small, but the closest we can come to really understanding this concept is by imagining $\frac{1}{2}$, $\frac{1}{4}$, $\frac{1}{8}$, and so on, diminishing like the inherited honors cited in Franklin's genetic argument. This is only a mathematical snapshot of a wide-ranging work of more general philosophy, but clearly we have strayed far from the mercantile concerns of Cocker's *Arithmetick*. Whereas the latter book is concerned with measuring spatial dimensions, amounts, and lengths of time, Locke considers what these concepts mean in the first place. He writes on reason, logic, morals, and the nature of ideas, but the concept of *number* figures prominently in many of his philosophical arguments.

Around the same time, young Benjamin also read *Logic; or, The Art of Thinking* (commonly known as the "Port Royal Logic") by Arnauld and Nicole.[43] These French Jansenists—theological allies of Pascal, for a time—were also contemporaries of Locke, and they share some of his notions.[44] For example, they note that the mind can perceive small numbers directly, but larger ones require some sort of conceptual leap. This is illustrated as follows. One can easily imagine a triangle, and with sufficient reckoning it is possible to determine that its angles sum to 180 degrees (equal to two right angles). However, one cannot literally picture a thousand-sided polygon, though we know that such a thing must exist, and we can even deduce that its angle sum must be equal to 1,996 right angles. What passes for a mental image of a thousand-sided figure could just as well stand for a *ten*-thousand-sided polygon, for there is nothing conceptually specific about it. (Locke makes the same point using different numbers.[45])

$$\underbrace{\qquad\qquad}_{\frac{1}{2}} \quad \underbrace{\quad}_{+\;\frac{1}{4}} \quad \underbrace{\;}_{+\;\frac{1}{8}} \quad \cdots \; = 1$$

Fig 2.11.

The Port Royalists also consider the nature of infinity. Here we learn that one-half, and one-half of one-half, and one-half of that, and so on *ad infinitum*, adds up to 1 (see figure 2.11); that one-third, plus one-third of one-third, and so on, adds up to one-half. (Such sums are known as *infinite series*.) Yet one-half plus one-third plus one-fourth, and so on, extends without bound.

Like Locke's *Essay*, this is a work of philosophy, not of mathematics—in fact, the *Logic* is particularly critical of earlier geometers—but the illustrations therein are often mathematical: the Pythagorean theorem, the definition of number. The authors reason from axioms, compare definitions (every rectangle is a parallelogram, but the converse is not true), classify triangles, and reference the geometric reasoning of Euclid and Descartes. These books so impressed Franklin that both *The Art of Thinking* and a collection of Locke's works were among the handful of books he contributed to the Library Company precisely ten years later.[46] Locke's broader influence on America's founders requires no comment.

This sixteen-year-old's reading list was quite impressive. Though Ben's dreams of a life at sea had been dashed, nevertheless he worked through two technical books on navigation "and became acquainted with the little geometry they contain." His recollection is overly modest, for there was considerable geometry in these two texts.[47] (We may forgive the incomplete remembrance; after all, that part of the memoir was jotted down nearly fifty years after the fact.) For example, in Captain Samuel Sturmy's *The Mariners Magazine*, one finds rather extensive instructions for straightedge-and-compass constructions, some theorems from Euclid (albeit without proof), trigonometry on the plane and the sphere, and hundreds of pages detailing applications to surveying and other problems of measurement. We can learn to "find the quantity of liquor in a cask that is part full" and to predict when the sun will rise and set. Also included are exhaustive descriptions of mathematical instruments, of which Franklin would amass a considerable assortment over his lifetime. As Sturmy explains, the "principal hand-maids" of a skilled seafarer are "arithmetick, geometry, trigonometry, and astronomy."[48] Though Franklin would never captain a ship, these two books must have provided some training that would prove useful for a writer of almanacs.

When not educating himself through reading, Ben was now ghostwriting for his brother's newspaper. The *New-England Courant*, one of the earliest newspapers to appear in the colonies, ran a series of fourteen letters by a moralizing widow named "Silence Dogood." Her correspondence arrived under the door of the printing house regularly, but unbeknownst to James, these amusing missives were written in the disguised hand of his younger brother. The Dogood letters comprise the earliest surviving writings by Benjamin Franklin. One of the more serious installments contains a proposal for an insurance scheme to support poor widows, a plan whose plausibility is buttressed by mathematical arguments based on Sir William Petty's *Political Arithmetick*.[49] Already in 1722, Franklin was making predictions based on population statistics, nearly thirty years before his *Observations Concerning the Increase of Mankind*.

The *Courant* was lively, innovative and independent, and therefore predestined for trouble in this Puritan stronghold.[50] Having run afoul of the authorities one time too many, James found himself in jail, and Ben became the de facto editor of the paper for a time. Imprisonment proved to be only a temporary inconvenience, but a second arrest warrant in 1723 for further offenses sent James into hiding, and the apprentice once again was in charge. If the experience provided invaluable training for America's first great newspaperman, it also gave him a taste of freedom, and the indenture would soon be severed. In fact, a court order forbade James from continuing to publish the newspaper without prior government approval (that is, censorship); interpreting that order as narrowly as possible, the brothers agreed that the *Courant* should simply continue under Ben's name instead. It was a hollow victory for the apprentice. Though the plan required his emancipation from the original indenture, a secret agreement was drawn up which required him to fulfill the remaining four years of servitude anyway. Meanwhile his master was free to administer the usual beatings to his underling, common enough in such relationships.

Tired of physical mistreatment and other injustices, Ben broke his contract and fled Boston. He sailed to New York, having sold some of his books to raise the funds. There he attempted to gain employment with William Bradford, the only printer in the city. There was no work to be had, but Bradford recommended his son's printing house in Philadelphia. (There was an opening there due to the early death of employee Aquila Rose, one of Philadelphia's first poets, who had served as clerk to the Pennsylvania House—a position Franklin would also hold.) The young runaway traveled south, mainly on foot, but on his arrival found the position filled. Instead he was directed to Samuel Keimer's printing house, where he finally found employment. It was readily clear to Franklin that both Keimer and his rival Andrew Bradford possessed talents inferior to his own. He would not remain a mere journeyman for long.

In 1724, Franklin sailed for London, having been promised letters of introduction and credit from the Pennsylvania governor, on whom he had made quite a favorable impression. Unfortunately the governor's letters made no impression on anyone, since he neglected to send them, and neither credit nor opportunity would be forthcoming. Not only was Franklin unable to obtain printing supplies as he had intended, but he could not afford the voyage back to the colonies. Two years passed, during which he worked for a London printer to earn his return fare. During this time, he wrote *A Dissertation on Liberty and Necessity, Pleasure and Pain*, which builds from axioms through a chain of logical propositions. Although it deals with the four ill-defined concepts in the title, the flavor is of an argument in natural philosophy; he even compares liberty to gravitation. The pamphlet's minor success led to an enlarged circle of acquaintances, including Henry Pemberton, an associate of Isaac Newton. According to Franklin's memoir,

"Dr. Pemberton . . . promised to give me an opportunity, some time or other, of seeing Sir Isaac Newton, of which I was extremely desirous; but this never happened."[51] It is tempting to imagine what might have transpired had they met, for the electrical theory advanced by an older Franklin would be modeled on the style of Newton's *Opticks*.[52] Had their paths intersected . . . but it is useless to speculate.

Finally, in 1726, Franklin made his way back to Philadelphia. After a brief interval working for a Quaker merchant (during which time he "studied accounts"), he returned to Keimer's employ. Two years later, he was able to establish his own printing business. During this time, he organized a club for mutual improvement called the *Junto*, a precursor to the American Philosophical Society (APS). Its earliest members were young tradesmen, some of them serious students of mathematics. They met to discuss topics in all areas of human endeavor, and to share thoughts, experiences, books, and the camaraderie that such academic fraternity provides.

Meanwhile, Franklin had decided that Philadelphia was in need of a second newspaper, one that would compete with Andrew Bradford's *The American Weekly Mercury*.[53] The plan was preempted when his idea was stolen by his former employer Samuel Keimer, and in 1728 the *Pennsylvania Gazette* was born. (Franklin had made the mistake of discussing his plan with one of Keimer's employees. In the words of Poor Richard, "Three may keep a secret, if two of them are dead.") The market would not now support a third weekly, but Franklin was not defeated. He penned anonymous letters that were printed in the *Mercury*, scathingly disparaging the *Gazette* and its publisher. The campaign paid off, for Keimer was forced to sell to his former assistant, and Franklin would turn the *Gazette* into the most successful newspaper in the colonies.

In 1730, Franklin's first child, William, was born, and later that same year Benjamin entered a common-law marriage with Deborah Read, who was most likely not the child's mother. They would have two other children. In short order he cofounded the Library Company of Philadelphia, the first subscription library in America (1731), and embarked on his career as the author of America's most important almanac (1732). All of these events are abbreviated severely here, for our purpose is not to create yet another complete biography of Franklin, but rather to consider his mathematical life. And as far as we can tell, it was indeed in the 1730s that Franklin began to draw his magic squares and circles. But just why did he do so? And why then?

That Franklin spent a great deal of time experimenting with magic squares is clear from the complexity of the matrices he left to us. Nevertheless, in all the vast literature on Franklin—thousands of books, and tens of thousands of articles—there is no real indication of what instigated those mathematical activities, apart from the occasional (erroneous) assertion that he was initially inspired by the works of Michael Stifel and Bernard Frénicle de Bessy; but as Franklin himself makes clear in a letter, he did not encounter either author until long after his experiments with magic squares had begun.[54] The whole affair has remained a mystery. Even in the massive *Papers of Benjamin Franklin* (thirty-eight volumes to date), there is nothing to suggest what sources originally inspired him.

The question has always fascinated me, because at that time the magic square was not yet well known in Britain, and even less so in the colonies.[55] The concept is absent from both Hodder's and Cocker's arithmetic primers. In any case, the frequent claims that Franklin's magical proficiency was a boyhood talent are entirely without foundation, so it would probably be useless to explore his early reading list for clues.[56] We must look elsewhere to find Franklin's inspiration.

Typically, those seeking to engage in analysis of historical figures and the influences upon them often focus on the personal library. For example, an index of the books owned by John Locke or Adam Smith or Isaac Newton is readily available for researchers to examine and dissect. In our case, that strategy poses some problems. After Franklin's death, his prodigious library—at 4000+ volumes by 1790, the largest private collection this side of the Atlantic—was dispersed, and no catalog was preserved. Complicating matters further, Franklin did not use a bookplate, unlike George Washington or Paul Revere, or, for that matter, Albrecht Dürer.[57] Even so, much of the library was painstakingly reconstructed in the twentieth century by bibliographic detective Edwin Wolf 2nd, longtime Librarian of the prestigious Library Company of Philadelphia.[58] My own exhaustive (and exhausting) search of this unpublished catalog, a card index thousands of titles strong, reveals few mathematics books. And those few were, for the most part, acquired too late in Franklin's life to have inspired or influenced his creation of the magic squares.[59] Thus we cannot know with certainty what the earliest influences were. Perhaps it was a book borrowed but never owned. The meme might even have been transmitted orally, or in correspondence no longer extant, leaving no trace for historians to ponder. Moreover, the catalog is far from complete, with three thousand volumes not accounted for.[60] Nevertheless, after several years of enthusiastic (some would say fanatical) investigations, I believe I have identified Franklin's sources.

One of these is the 1708 English edition of Ozanam's *Recreations Mathematical and Physical.*[61] First published in the 1690s in France, then revised by a variety of editors over the next 150 years, Ozanam would remain the most important reference on recreational mathematics for two centuries.[62] This early translation, condensed to a single volume, contains enjoyable diversions in not only arithmetic and geometry but also physics, astronomy, and other scientific areas. There are card tricks, number games, and such topics as: How to shoot a pistol behind one's back using a mirror. How to build a scale that only *appears* to be honest. How to make gunpowder. Though the complete text would make for a lengthy course of study, one need not read very far before encountering magic squares.[63] The reason this specific influence has remained unidentified is simply that the supporting documents cited below have never been included in any printed collection of Franklin's papers to date.

Why do I think this book helped send Franklin on his mathematical odyssey? First, a document written in Franklin's hand, dating to the mid-1730s, lists books intended for his Library Company, among them "Ozanam's Mathematical Recreations."[64] Bear in mind that Franklin's tenure as clerk of the Pennsylvania Assembly commenced in 1736, and that his clerkship is the earliest time when we can be sure that he experimented with magic squares; thus the timing is just right. Second, the minutes of the Company from late 1734 show that the directors (Franklin and nine others) ordered a copy of the *Recreations*, and another entry five months later acknowledges that it has arrived safely from England.[65] At a time when the library held barely 200 volumes—a trifling amount of linear shelf space—one could hardly fail to notice a particular title, especially if one were an active member (founder, subscriber, and variously director, librarian, and secretary) who spent so much time perusing its holdings.[66] For Franklin writes: "This library afforded me the means of improvement by constant study, for which I set apart an hour or two each day; and thus repaired in some degree the loss of the learned education my father once intended for me."[67] Edwin Wolf believed that Franklin authored an anonymous letter to the *Pennsylvania Gazette* which states: "When my daily labour is over, instead of going to the alehouse, I amuse myself with the books of the Library Company, of which I am an unworthy member."[68] Someone who passed the hours after work browsing the collection would certainly have run across the *Recreations*, which in any case was less daunting than Ozanam's five-volume mathematics course, also on the shelf. Moreover, a cursory browse is all it would take to happen upon the ninth of nearly three hundred topics addressed by Ozanam. Finally, Franklin went to the trouble of obtaining a larger French edition much later in life, a fact which has been obscured by his assistant's careless abbreviation of the title (and complete deletion of the author's name) when cataloging the book.[69]

It was also in the 1730s that Franklin began to author an annual pamphlet, thereby becoming the most celebrated almanacist since Nostradamus. To assist in the writing of *Poor Richard*, Franklin gathered an impressive array of English almanacs, some dating back as far as the 1660s, on which to base his own. These were later bound together in annual volumes. In 2004, on a visit to the American Philosophical Society, I met Roy Goodman, curator of printed materials, whose encyclopedic knowledge of Ben Franklin far exceeds my meager seven-year study. Goodman pointed out that, unlike much of Franklin's library, the almanac collection survives largely intact, and it is housed today in the APS library. It turns out that this collection includes a fairly complete set of the earliest decades of *The Ladies Diary*, a mathematical periodical of great historical significance.

John Tipper's *The Ladies Diary, or, the Woman's Almanack* was a fantastically successful British annual that combined the typical calendar almanac with challenging riddles and mathematical puzzles aimed at the "fairer sex." Considering the extent to which women have been marginalized in the sciences throughout history, it is remarkable that the eighteenth-century *Diary* both invited solutions from women and regularly acknowledged the solvers. Perhaps Franklin's progressive views on the education of the sexes may be traced to the vehement editorializing of Tipper's almanac.

Tipper, a Coventry schoolmaster, inaugurated his almanac in 1704. Franklin's own collection of this title runs from 1706, the year of his birth, through 1752, except for three years in which the entire volume is absent. In those instances, every other title collected by Franklin is likewise missing, from Moore's *Vox Stellarum* to Andrews's *Remarkable News From the Stars*. Thus, several bound volumes were separated from the rest of the collection long ago, among them the almanacs for 1727, which explains why our second mathematical source has gone unnoticed.

By this time, the editorship of the *Diary* had passed to Henry Beighton, a mathematician and engineer who worked on an early version of the steam engine, long before the arrival of James Watt.[70] Under "New Arithmetical Questions" we find a little poem:

> 'TIS to you, Lovely Ladies, I sue and submit,
> (Who outvie *Sidrophel* in Magic and Wit)[71]
> For Solution of this Knotty Problem propos'd,
> By which Undertaking my Senses are doz'd;
> To find by which Canon the Squares you do fill
> Which are Magical call'd, and by that try your Skill,
> To place all these Numbers* so that the Amount;
> Just half a score ways Seventy four you may count:
> If you'll answer but this, now your self do assure,
> I will meddle with what they call *Magic* no more.

* 8,9,10,11,14,15,16,17,20,21,22,23,26,27,28,29

The *LADIES Diary*,
OR, THE
Woman's ALMANACK,
For the YEAR of our LORD, 1728.
Being *Biſſextile*, or LEAP-YEAR:
Containing many Delightful and Entertaining *Particulars*,
Peculiarly Adapted for the *Uſe* and *Diverſion* of the
FAIR-SEX.

Being the Twenty Fifth ALMANACK ever Publiſh'd of that kind.

1. HAIL! happy *LADIES* of the *BRITISH* Iſle.
   On whom the GRACES and the MUSES ſmile.

2. LONG had your lovely *Shape*, and matchleſs *Mein*,
   The Wonder of our Neighbring Nations been;

3. NATURE to make your *Triumph* more compleat,
   To peerleſs CHARMS has added piercing Wit.

4. NO more let *SCYTHIA* vaunt her FEMALE-HOST,
   Nor their SEMIRAMIS th' *Aſſyrians* boaſt:
   WIT joyn'd to BEAUTY, *Fame* ſhall now record;
   Which lead more Captive than the Conq'uring Sword.

Printed by *A. Wilde*, for the Company of *Stationers*, 1728.

Fig 2.12. A typical cover of *The Ladies Diary*, "for the use and diversion of the fair sex." American Philosophical Society Library.

This question was followed by another charming mathematical verse asking the reader to calculate the precise depth of the Great Flood. (Sometimes I wonder whether modern mathematics courses might be made more palatable through the use of some creative rhyme.)

The question, then, is this: Arrange the given numbers so that you get a total of seventy-four in ten different ways.[72] Ambiguous as stated, it can be rephrased thusly: How does one position those sixteen numbers in a 4 × 4 matrix so that all four rows across, all four columns down, and both diagonals add up equally? Even if Franklin never saw the 1727 question, he did own the 1728 solution key, which explains: "If the numbers as given be placed in four Rows, you need only let the Diagonals stand, and cross places with the other numbers in these squares [figure 2.13]." The caption below the right-hand figure essentially repeats the original question.

In the grid on the left of figure 2.13, the numbers are listed in order. Draw an **X** over the diagram, covering both diagonals; these entries are to be left alone. The other numbers not covered by the **X** will be dialed around 180 degrees. The result is the magic square on the right. This little trick is similar to the one employed previously by the alchemist Agrippa and the artist Albrecht Dürer, and for that matter to a technique later used by Ben Franklin. This, then, is another source that met with Franklin at just the right time. The *Diary's* importance might be gauged by the fact that it holds a

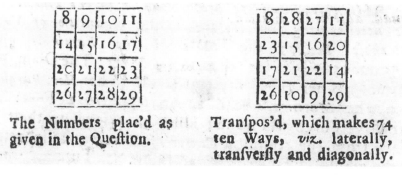

The Numbers plac'd as given in the Queſtion.

Tranſpos'd, which makes 74 ten Ways, *viz.* laterally, tranſverſly and diagonally.

Fig 2.13. The *Diary's* solution. American Philosophical Society Library.

prominent place at the front of Franklin's bound volumes, in around half of the years in which it appears. An earlier issue, probably acquired around the same time, held a stronger challenge:

> You that delight in figures, try your skill,
> A magic square with numbers for to fill;
> One to a hundred, numbers just must be,
> Which to the numbers of the squares agree.
> But farther, you must them so just contrive,
> Twenty-two ways, make five hundred and five.
> No two squares alike in numbers must be;
> But ten in breadth, and ten in length, let's see.

Since the 10×10 answer was printed without any explanation, this poetic problem was likely not as helpful to aspiring magicians.[73]

Challenge

We have seen that the *Sagrada Família* square (figure 2.10) can be obtained from a rotated Dürer's square by subtracting a magic square composed only of zeroes and ones, which we may call *M* for convenience. Show how the 4×4 magic square given in the *Ladies Diary* can be obtained from Agrippa's Jupiter square, by a similar (if slightly more complicated) process. *Hint*: Use some of the eight rotations and reflections of *M*, as well as a constant matrix of 7s.

(For solutions to boxed problems, see appendix 5.)

It is impossible to determine which of these two sources exerted the greater influence. Beighton's *Diary* was published first, though that does not mean that Franklin acquired it first. Ozanam's *Recreations* fills the squares with consecutive numbers, as Franklin did, but illustrates squares whose dimensions are odd numbers. On the other hand, the *Diary* square has even dimensions, as Franklin's squares inevitably did, but the entries are not consecutive. He appears to have synthesized the two sources before taking the magic square to ever greater heights.

There may be other influences that we have missed. Consider a 1729 letter to *The American Weekly Mercury*, pseudonymously authored by Franklin, shortly before he acquired the *Pennsylvania Gazette*. The epistle from "Busy Body," a successor to Silence Dogood, quotes the equally fictional astrologer Titan Pleiades, who bolsters his occult credentials by claiming to have read "Cornelius Agrippa above 300 times."[74] Might we assume that Franklin read Agrippa at least once? Did he see the magic squares there? It is probably a stretch to say so.[75]

One final connection is equally tenuous, if more intriguing. Like many of the Founding Fathers, Benjamin Franklin was a prominent figure in colonial Freemasonry. He printed the first Masonic book to be published in the New World. Later in life he would be received by lodges in Scotland and France.[76]

Masonic culture inherits rites and symbols from many mystical traditions. The objects which populate Dürer's *Melencolia* allegedly represent symbols of Freemasonry. The exception is his magic square, but is it really an exception? So much secrecy surrounds Masonic traditions that it would come as no surprise if this symbol, too, were somehow allied with the "Craft." After all, it seems illogical for every allusion *but one* to be in agreement in Dürer's masterpiece.

A well-circulated story, albeit probably a fabrication, is relevant here. When the Prophet Joseph Smith, founder of the Church of Jesus Christ of Latter-day Saints, was martyred at Carthage Jail in 1844, it is said that he carried a Jupiter medallion on his person; this is a silver talisman inscribed with Agrippa's 4×4 magic square. What concerns us is not the veracity of the story—it appears to be contrived in order to denigrate Smith by associating him with a superstitious practice—but rather the fact that in most versions of the story, this talisman is described as a "Masonic jewel" (a term which more typically refers to something quite different). Thus the magic square and Freemasonry are allied in a popular legend, though one of dubious authenticity.

Moreover, Freemasonry claims descent from the Knights of the Temple of Solomon. When the Order of the Temple was dissolved by Pope Clement V in the fourteenth century, a long list of spurious charges was leveled at the Crusaders: worship of idols, subversion of sacred rituals, and many more of the usual heresies. (The knights showed considerable fortitude under questioning. Torture, or simply the threat of torture, is an efficient method of evoking confessions from guilty and innocent alike, yet many of the brethren refused to admit wrongdoing even as they were burned to death.[77]) But they were also accused of having been tainted by knowledge of sorcery, through their contact with the enemy. If, indeed, the Templars acquired mystical knowledge of medieval Islamic numerology, might the numerology associated with magic squares be part of the Freemasons' cultural inheritance? As far-fetched as it sounds, such speculations are bolstered by the tale of the Crusaders purchasing magic square plates in Jerusalem. Without entering into the massive literature on Masonic history, much of it of dubious historicity (holy grails, Christ's secret children, and so forth), we may at least ask whether the magic square has a place in Masonic ritual.

With this in mind, I spent many hours researching that very question at the Masonic Library and Museum of Pennsylvania, but came up empty until I found a concrete reference to the seven planetary squares. According to a prominent Masonic historian, "these magic squares and their values have been used in some of the high degrees of Masonry."[78] The precise context of this purported use is not revealed. Were "high degrees" conferred on Franklin? Quite so, for in 1734 he was elected Grand Master of the Grand Lodge of Pennsylvania (he later ascended to Provincial Grand Master), and this event immediately preceded his initial experimentation with magic squares.[79]

Thus, we have identified two reliable influences and two entirely hypothetical ones: a book, an almanac, an alchemist's guide, and a Masonic ritual that might have been used at St. John's Lodge in Philadelphia. By coincidence, each of these falls within the same general time frame, just prior to Franklin's clerkship in the Pennsylvania Assembly. Absent a smoking gun, it is probably the most we

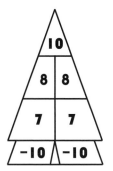 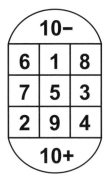 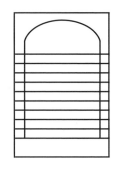

Fig 2.14. Shuffleboard.

can hope for. This represents, I think, sufficient inquiry for a question that is likely never to be resolved definitively.

You don't need a Jupiter medallion to try your luck at the game of *shove-groat*; an ordinary coin will do. This ancestor to modern shuffleboard, which today is played on a table or floor, was once known by many names, such as *shovel-groat* or *shovelboard*, and it was remarkably popular in England for hundreds of years; in the opening scene of Shakespeare's *The Merry Wives of Windsor*, a chap named Slender complains that he has been robbed of "seven groats in mill-sixpences, and two Edward shovel-boards that cost me two shilling and two pence apiece." To play, slide a coin down a tabletop. Scoring is based on where the coin comes to rest, according to some predetermined set of rules for your particular version of shove-groat. (A groat is worth four pence, or one-third of a shilling; currency conversion is useful not just in trade but also in tavern games.) Three different scoring boards are shown in figure 2.14. The first of these is most familiar to shipboard travelers today. The second configuration is also used; notice that there is a magic square hiding in plain sight! Does this mean that Ben Franklin might have been inspired by a popular English diversion? No, because in his day the board used was the unmarked one at right.[80] I mention this example not only to preempt a possible theory of influence, but also to underscore once again how the magic square has permeated the popular consciousness in so many strange and unexpected ways.

Notes

1. A justifiably skeptical discussion of this earlier dating is given by Paul Carus in "Reflections on Magic Squares" (p. 122), in W. S. Andrews's classic *Magic Squares and Cubes*, 2nd ed., Chicago: The Open Court Publishing Company, 1917.

2. Also: *luo shu, loh shu*. In a sense, the lo shu turtle really did carry the world on his back, or at least a symbolic model of the universe.

3. Xueqin Li, Garman Harbottle, Juzhong Zhang, and Changsui Wang, "The earliest writing? Sign use in the seventh millennium BC at Jiahu, Henan Province, China," *Antiquity*, Vol. 77, No. 295, 2003, pp. 31–44. For statements pro and con, see Paul Rincon, "'Earliest writing' found in China," BBC Science, April 17, 2003, http://news.bbc.co.uk/2/hi/science/nature/2956925.stm. The interpretation of this artifact is controversial, but even if ultimately rejected it is still true that "the oldest known examples of Chinese writing" are found on "fragments of tortoise shells and deer scapulae," according to Georges Jean, *Writing: The Story of Alphabets and Scripts*, New York: Harry N. Abrams, 1992. See also T. R. Tregear, *The Chinese*, New York: Praeger, 1973.

4. Jacques Sesiano, *"Quadratus Mirabilis,"* in *The Enterprise of Science in Islam*, edited by Jan P. Hogendijk and Abdelhamid I. Sabra, Cambridge: The MIT Press, 2003, pp. 199–234.

5. Some of the earlier texts are described in Schuyler Cammann, "The Evolution of Magic Squares in China," *Journal of the American Oriental Society*, Vol. 80, No. 2, 1960, pp. 116–24. See also George Gheverghese Joseph, *The Crest of the Peacock: Non-European Roots of Mathematics* (new edition), Princeton: Princeton University Press, 2000; and Joseph Needham, *Science and Civilisation in China, vol. 3: Mathematics and the Sciences of the Heavens and the Earth*, New York: Cambridge University Press, 1959.

6. Schuyler Cammann, "The Magic Square of Three in Old Chinese Philosophy and Religion," *History of Religions*, Vol. 1, No. 1, 1961, pp. 37–80; Cammann, "Islamic and Indian Magic Squares, Part I," *History of Religions*, Vol. 8, No. 3, 1969, pp. 181–209, and "Part II," Vol. 8, No. 4, 1969, pp. 271–299; Jacques Sesiano, "Magic Squares for Daily Life," in *Studies in the History of the Exact Sciences in Honour of David Pingree*, editors Charles Burnett et al., *Islamic Philosophy and Science: Texts and Studies* Vol. 54, 2004, pp. 715–734; Walter William Skeat, *Malay Magic: Being an Introduction to the Folklore and Popular Religion of the Malay Peninsula*, London: Macmillan and Co., 1900; Sesiano, 2003; Joseph, 2000; Cammann, 1960. It has even been suggested that the line of transmission worked the other way, with the original invention beginning in what is now the Arab world and only thence going to China.

7. Susan Blackmore, *The Meme Machine*, New York, Oxford: Oxford University Press, 1999. I should note that Martin Gardner, the modern high priest of magic squares, has objected to the very notion of a meme. (Robert Aunger, ed., *Darwinizing Culture: The Status of Memetics as a Science*, New York: Oxford University Press, 2000, p. 2.) I suspect that he would be unhappy with my identification of a magic square as a meme.

8. Our triangle follows the definition used in Helmut Krämer, "Magische Dreiecke, endliche Gruppen und Kombinatorik," *Mitteilungen der Mathematischen Gesellschaft in Hamburg*, Vol. 19, 2000, pp. 155–172. Some variety of magic triangle was studied in India by the mid-1300s; see Parmanand Singh, "Total Number of Perfect Magic Squares: Nārāyana's Rule," *Mathematics Education (Siwan)*, Vol. 16, No. 2, 1982, pp. 32–37. Our hexagram is from Harold Reiter and David Ritchie, "A Complete Solution to the Magic Hexagram Problem," *The College Mathematics Journal*, Vol. 20, No. 4, 1989, pp. 307–316.

9. Camman, 1969, Part I.

10. Here is the completed matrix:

11	24	7	20	3
17	5	13	21	9
23	6	19	2	15
4	12	25	8	16
10	18	1	14	22

11. Other equivalent names are *panmagic*, *Nasik* square.

12. Cammann, 1969, Parts I and II, and Sesiano, 2004.

13. This is attributed to a manuscript titled *Lutfī'l Maqtūl*, which may have conflated the magic *square* with the Delian *cube*.

14. The purposes described throughout this paragraph are from Sesiano, 2004.

15. The invention of the magic square has been attributed variously to China, Islam, and India, but many sources of the eighteenth and nineteenth centuries claim an Egyptian and/or Pythagorean pedigree. Their assertion is without foundation.

16. Warning: These claims have not been evaluated by the Food and Drug Administration.

17. "Manuel Moschopoulos's Treatise on Magic Squares," translated by John Calvin McCoy, *Scripta Mathematica*, Vol. 8, No. 1, 1941, pp. 15–26. (See especially his first endnote regarding discrepancies in dating.) Also P. G. Brown, "The Magic Squares of Manuel Moschopoulos," *Convergence*, Vol. 1, 2005, http://mathdl.maa.org/convergence/1/.

18. Sesiano, 2004. See also Karl Anton Nowotny, "The Construction of Certain Seals and Characters in the Work of Agrippa of Nettesheim," *Journal of the Warburg and Courtauld Institutes*, Vol. 12, 1949, pp. 46–57. The twelfth-century philosopher-astronomer Abraham ibn Ezra is sometimes cited as having delivered the magic square to Europe, but if so, his influence in this regard took some time to take hold.

19. Simon de la Loubère, *A New Historical Relation of the Kingdom of Siam* (English Translation), London: Printed by F. L. for Tho. Horne, 1693.

20. Nowotny, 1949. Many authors claim that the concept of a magic square passed from Moschopoulos to Trithemius to Agrippa; the second step in that chain is well established, but it seems to me that the first step is pure

speculation. Ibn Ezra is another possible influence, as is Agrippa's near-contemporary Luca Pacioli.

21. I have used the reprinting of *De occulta philosophica* as it appeared in Agrippa's *Opera* (1600). English translations are currently available both in print and on the internet.

22. This is from Hutton's 1795 *Dictionary*, about which I will say more in chapter 6.

23. It is often claimed that the oldest known instance of the inscription appears on a wall in Pompeii, but I have not yet found a reliable source to support that assertion. There are many interpretations of the SATOR square, both straightforward and scrambled, but these are for the most part unconvincing.

A similar palindrome, far more obscure, may date back to the late twelfth century. It reads "Natas areda tedet adera Satan." (British Library manuscripts 8 F. IX: Theological tracts, in Latin, and, bound with them, part of the French metrical romance of Guy of Warwick.)

24. An entire book could be written on the mathematical side of Albrecht Dürer, but there is no room for a lengthy discussion here.

25. The latter fact was brought to my attention by Todd Pelletier.

26. Josep Maria Carandell, *El Temple de la Sagrada Família*, 1997, pp. 70–79, and Paul C. Pasles, "Some Magic Squares of Distinction," *Math Horizons*, Feb. 2004, pp. 10–12.

27. An earlier work of fiction was Paul Calter's *Magic Squares* (1977). One composer who has examined the magic square in greater detail than Cage is Peter Maxwell Davies.

28. Cardano's comments on magic squares are in *Practica arithmetice & Mensurandi singularis* (1539), where he explains a method of construction similar to the one that appears in figure 5.4.

29. Jacques Darriulat, *L'arithmétique de la grâce: Pascal et les carrés magiques,* Paris: Les Belles Lettres, 1994. In the ensuing centuries, a host of brilliant mathematicians, primarily known for other work, would devote themselves to the academic study of magic squares; these include names instantly recognizable to any serious student of mathematics: Fermat, Euler, Ramanujan, Bieberbach, D. N. Lehmer.

30. François-Guillaume Poignard, *Sur les Quarrés Magiques* (also known as *Traité des Quarrés Sublimes*), Brussels, 1704; also Hutton's *Dictionary*, which we will encounter in chapter 6.

31. Most likely this was the same George Brownell who taught in Philadelphia in the 1720s and 1730s. (Around the time his advertisements disappeared from the Boston newspapers, they began to appear in the *American Weekly Mercury* and the *Pennsylvania Gazette*.)

32. Details of Franklin's life up to this point are taken from the *Autobiography*.

33. *Hodder's arithmetick: or, That necessary art made most easy. Being explained in a way familiar to the capacity of any that desire to learn it in a little time. By James Hodder, writing-master. The five and twentieth edition, revised,*

augmented, and above a thousand faults amended, by Henry Mose, late servant and successor to the author, Boston: Printed by J. Franklin for S. Phillips and others, 1719. This American *Hodder's* was preceded in New York and Philadelphia by *The Young Man's Companion*—later retitled *The Secretary's Guide*—of which only one part was devoted to "Arithmetick made easie." (The *Companion* was printed by William and Andrew Bradford, whom we shall meet later on in this chapter.) See John Alfred Nietz, *Old textbooks . . . from Colonial days to 1900*, Pittsburgh: University of Pittsburgh Press, 1961.

34. *Autobiography*.

35. This according to dozens of advertisements in *The Pennsylvania Gazette* (for instance March 22, 1739; May 21, 1741; Oct. 2, 1746).

36. "To My Name 1713," Commonplace Book of Benjamin Franklin the Elder, in *Papers*, 1959, Vol. 1, pp. 5–6.

37. M. Schele De Vere, *Americanisms: The English of the New World*, New York: Charles Scribner & Company, 1872; also Ebenezer Cobham Brewer, *Dictionary of Phrase and Fable*, London: Cassell and Company, 1898. A more recent American version of this expression is the phrase "according to Hoyle," with card games taking the place of arithmetic as the touchstone for rule following.

38. The exercise is question 30, chapter 10 in most editions. One approach to its solution would be to put all of the quantities in terms of the smallest unit, the penny (currency) and the pound (weight, or more properly mass). The first chest carries $17(112) + 3(28) + 14 = 2002$ pounds of sugar at 560 pence per hundredweight ($560/112 = 5$ pence per pound). The second chest carries 2065 pounds of sugar priced at $4\frac{1}{2}$ pence per pound. The mixture can be valued using a weighted average: $2002(5) + 2065(4\frac{1}{2})$ divided by $2002 + 2065$ equals $19302.5/2067$ pence per pound. A hundredweight of the mixture is worth 112 times that rate, which works out to 531 47/83 pence, or just over 2 pounds 4 shillings $3\frac{1}{2}$ pence. Cocker's various editions give three different values for the fractional remainder, each time incorrectly.

39. Specifically, he implies in the *Autobiography* that the drawing of magic squares and circles was simply a cure for boredom, and in a letter to Peter Collinson (see chapter 5) he says that his time could have been spent "more usefully."

40. The quotations in this chapter are from the first and third editions of *An Essay Concerning Humane* [sic] *Understanding*. (Today it is seldom referenced under the archaic spelling.)

41. A European encounter documenting this claim is Jean de Léry, *Histoire d'un Voyage Fait en la Terre du Brasil* (1570).

42. Both the spelling and the meaning of these terms have varied.

43. Antoine Arnauld and Pierre Nicole, *Logic; or, The Art of Thinking* (English translation), 2nd edition, London: Printed by T. B. for Randal Taylor, 1693. There is no evidence that Franklin read Arnauld's *Nouveaux Éléments de Géometrie*.

44. Not only theological allies, they also rely on a geometry manuscript written by Pascal.

45. Locke contrasts a 999-sided figure with one of a thousand sides. He calls the latter a *Chiliaëdron*. (A *chiliad* refers to a set of one thousand items. For example, Cotton Mather refers to millenniarian belief as the "doctrine of the chiliad.")

46. Library Company minutes, Vol. I, 1731–1768.

47. The quotation is from the *Autobiography*, in which Franklin identifies "Seller's and Sturmy's books of navigation." The latter must refer to Samuel Sturmy's *The Mariners Magazine*, and the former to either *Practical Navigation* or possibly *An Epitome of the Art of Navigation* by John Seller.

48. Capt. Samuel Sturmy, *The Mariners Magazine, Stor'd with These Mathematical Arts . . .*, 2nd edition, London: Printed by Anne Godbid, for William Fisher et al, 1679. (In many editions of the *Autobiography*, this author's name is mistakenly given as "Shermy.")

49. The letter was published August 13, 1722. Most of it is purportedly taken verbatim from another source, so that Petty is only a secondary source, but I do not know if that was only a device to confer respectability on an argument that was actually composed by Benjamin/Silence.

50. This would become something of a family tradition. Two generations later, Ben's grandson B. F. Bache would be imprisoned for publishing the *Aurora*, a newspaper critical of President Adams.

51. Pemberton edited the third edition of Newton's *Principia*. "Pemberton's View of Sir Isaac Newton's Philosophy" appears on a list of books sold by Franklin's printing house in 1744.

52. I. Bernard Cohen, *Franklin and Newton; an Inquiry into Speculative Newtonian Experimental Science and Franklin's Work in Electricity as an Example Thereof*, Memoirs of the American Philosophical Society, Vol. 43, published for the APS by Harvard University Press, Cambridge, 1966.

53. In 2006, both major Philadelphia dailies are owned by a single organization. Where are the Ben Franklins of today?

54. We examine that letter in detail in chapter 5.

55. Early English Books Online (EEBO) serves as a reliable barometer for Britain until 1700, and the "Evans" database for America through the eighteenth century. The only reference to the magic square in EEBO is in an English translation of de la Loubère (1690s), while Evans shows no American appearance before 1799. Among English publications after 1700, the first are *Lexicon Technicum*, 1704, and a translation of Ozanam's *Recreations*, 1708. (Daniel Defoe cites the former reference when describing a magic square in his *History of the Devil*, 1727, wondering "what nefarious operations are wrought by this concurrence of the numbers.") Other early appearances include Ephraim Chambers's *Cyclopædia* (1728) and *The Builder's Dictionary* (1734).

56. Candace Fleming (*Ben Franklin's Almanac: Being a True Account of the Good Gentleman's Life*, p. 47) states: "As a boy, Ben became obsessed with creating 'magic squares of squares'. . . . Simple at first, his magic squares

grew more complicated as his math skills grew." H. W. Brands (*The First American*, p. 207), claims that "Franklin had encountered magic squares as a boy," probably basing the assertion on Van Doren, another secondary source. In fact, there is no primary source evidence at all that Franklin met the magic square prior to adulthood, much less experimented with his own. It is true that he once recalled having made magic squares in his "younger days" (chapter 5), but he was around forty-five years old when he wrote those words, so it is difficult to know exactly what he meant.

Most of Franklin's major biographers have made far more serious errors on the subject of the magic squares, as have the authors of children's literature on the topic. Among other failings, typically they give the impression that Franklin drew straight-diagonal squares only, which is rather like praising Edison for his deft use of candles: it completely ignores his signature improvements and credits him instead for the discoveries of the past.

57. See James Keenan, *The Art of the Bookplate*, New York: Barnes & Noble Books, 2003. That Franklin had no bookplate is explained in James Green, *Poor Richard's Books*, Philadelphia: Library Company of Philadelphia, 1990; and in earlier articles of Edwin Wolf.

58. Green, *Poor Richard's Books*.

59. The Wolf catalog of Franklin's books will finally be published in 2006, but it is not yet available at this writing.

60. Green, *Poor Richard's Books*.

61. Jacques Ozanam, *Recreations mathematical and physical; Laying down, and Solving Many Profitable and Delightful Problems*, London: Printed for R. Bonwick and others, 1708.

62. He was finally supplanted by W. W. Rouse Ball's *Mathematical Recreations and Problems of Past and Present Times* (1892), which later became *Mathematical Recreations and Essays*. See chapter 8.

63. The material begins on p. 33 of that 500-page volume.

64. J. A. Leo Lemay, *Benjamin Franklin: A Documentary History*, http://www.english.udel.edu/lemay/franklin/.

65. To be precise, the later minute explains that certain titles are omitted from the shipment, but Ozanam is not among those listed as missing (and it does appear in later catalogs). Library Company minutes, Nov. 13, 1734 and April 18, 1735.

66. The oldest surviving printed catalog is from 1741, but most of the volumes listed there can be placed in time by comparing the minutes for 1731–1741. I have based this estimate of the library's size on a page-by-page reading of those minutes.

67. *Autobiography*.

68. Introduction to *A Catalogue of Books Belonging to the Library Company of Philadelphia: A Facsimile of the Edition of 1741 Printed by Benjamin Franklin, with an Introduction by Edwin Wolf 2nd*, Philadelphia: Library Company of Philadelphia, 1956.

69. It is listed as "Recreations Phisiques Nouvelles 4 [volumes, no author]."

70. His contributions are detailed in L.T.C. Rolt, *Thomas Newcomen: The Prehistory of the Steam Engine*, Dawlish: David and Charles, 1963.

71. That is, an astrologer. Sidrophel is a character in Samuel Butler's satirical poem *Hudibras*, an unflattering portrayal based on the renowned seventeenth-century astrologer and almanac writer William Lilly. Lilly's almanac, *Merlinus Anglicus Junior*, continued after his death, and found its way into Franklin's personal collection of almanacs and ephemerides.

72. The exercise, question 126, is credited to "Mr. Tho. Dod." and the solution to one Elias Colbourn.

73. *Diary* for 1718, solved 1719.

74. Lemay's *Documentary History* attributes *The Busy Body No. 8* to Franklin. The astrologer's pseudonym is probably based on Titan Leeds, an almanac writer we will meet in chapter 3. (The Pleiades and the Titans probably require no further explanation.)

75. It seems unlikely that such a critical thinker as Franklin would bother with such a work. But then Isaac Newton was a dedicated alchemist, so who can say for sure?

Franklin might also have encountered these magic squares much later, in the 1600 edition of Agrippa's *Opera*, which was in James Logan's library.

76. Julius Friedrich Sachse, *Benjamin Franklin as a Free Mason*, Lancaster: The New Era Printing Company, 1906.

77. Piers Paul Read, *The Templars*, New York: St. Martin's Press, 1999.

78. Albert G. Mackey, *An Encyclopedia of Freemasonry and Its Kindred Sciences*, Vol. I, New York and London: The Masonic History Company, 1909. While the alphabetic SATOR matrix referred to earlier in this chapter is sometimes called the "templar magic square," and Freemasonry is traditionally linked to the Knights Templar, Mackey's *Encyclopedia* specifically refers to the numerical type of square.

79. There are many complicating factors in this line of reasoning. Masonic ritual has varied over time and place. There is also the matter of the *ancients* and the *moderns* and other minutiae beyond the scope of this book.

80. A simplified version was known in my childhood as "paper football," conveniently played by two opponents whenever the study hall proctor was distracted.

3 Almanacs and Assembly

All publicity is good, except an obituary notice.
—Brendan Behan

s publicity stunts go, faking one's own death is a time-worn but effective technique. Faking someone *else's* death in order to drum up business constitutes a more creative strategy. And that's just what Ben Franklin did, during the celebrated feud between Poor Richard and his professional rival Titan Leeds. What the historians have missed, however, is the fact that the opening salvo in this battle was written in the language of mathematics.

Leeds worked for the New York and Philadelphia Bradfords. In the spirit of *The Ladies Diary*, his almanac for 1731 posed three questions to the "Sons of Art"—that is, fellow astrologers. Readers were cautioned that "the answer sent to the above questions cannot be accepted without their solutions."[1] (As we say in math class, you must show your steps!)

Question I. There is a wall containing 18225 cubical feet. The height is five times the breadth, and length eight times the height. I demand the breadth of the wall?

Question II. Suppose a ship sails between the S. and W. till the sum of the difference of latitude and [difference of longitude] be 7, and her distance 5. I demand the difference of latitude and [longitude] severally? [Assume for simplicity that the earth is flat.]

> *Question* III. Suppose the area of an equilateral triangle to be 600. The
> sides are required?[2]

At this time Franklin was already a polymath, a student of many subjects; but not yet a self-proclaimed *philomath*, the traditional title assumed by almanac writers.[3] However, he did publish almanacs by other authors, such as local glassmaker/mathematician Thomas Godfrey, a Junto member and cofounder of the Library Company.[4] Godfrey quickly solved the questions posed by Titan Leeds. It would be reasonable to expect some acknowledgement in the next year's installment, as was standard practice in the *Diary*, and indeed he was so acknowledged. But next year was not soon enough for Godfrey's publisher, Benjamin Franklin. The *Pennsylvania Gazette* ran an advertisement for Godfrey's own almanac "containing the eclipses, lunations, judgment of the weather, the time of the sun's rising and setting, moon's rising and setting, . . . time of high-water, fairs, courts, and observable days," which could be purchased at Franklin's printing office.[5] The advertisement was accompanied by a rather unkind verbal assault leveled at Titan Leeds. It seems Godfrey had worked through the questions in under thirty minutes, "as the printer [Franklin] hereof can testify," suggesting that that they were too easy. Godfrey sent his solutions to Leeds along with some new questions of his own, according to the *Gazette*, but Leeds claimed not to have received them—though this was a mere "pretence." Godfrey sent his questions again, but now six weeks had passed, and still there was no acknowledgement. Thus "it is thought proper to publish the said questions, that there may be no room for further excuse. Note, a copy of the aforesaid solutions is in the hands of the printer." The implication, probably correct, is that Leeds was by far the inferior mathematician. Godfrey certainly invented harder questions. If you are versed in high school mathematics, the three problems above should pose only a little difficulty, but the same cannot be said for the counterchallenge:

> *Question* I. Suppose 2 ships being in the same latitude, distant from
> each other 100 leagues, sail directly north 500 leagues, (20 in a degree)
> and then are 70 leagues apart: What are the latitudes?

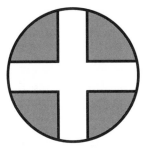

Fig. 3.1.

Question II. Suppose 2 roads of = breadth to cross each other at right-angles in the centre of a given circular piece of ground, and take up the half or third part of that circular piece of ground: What is the proportion of the breadth of the roads to the diameter of the circle? [See figure 3.1.]

Question III. Suppose a ship in north latitude, sailing between the north and east, makes her distance 50 minutes more than her diff. latitude, 40 minutes more than her departure, 30 minutes more than her diff. longitude. What is her course, distance, and both latitudes?[6]

The advertisement might have been placed by Godfrey alone, but it sounds like the work of Franklin. After all, he was not merely the printer of the newspaper. According to a foremost expert on Franklin, he "performed all operations—wrote copy, set type, pulled sheets, and trundled the papers through the streets. To a remarkable degree the *Gazette* was Franklin."[7] An editor of his papers says that "in its earliest years at least [the *Gazette*] was written by him from beginning to end."[8] Besides, this assault fits with his later treatment of Leeds.

The victim of the attack had already suffered a serious identity crisis. In the 1720s a curious thing happened: two Philadelphia printers both claimed to publish the true Leeds almanac. Samuel Keimer's pamphlet warned, "beware of the counterfeit one," and an announcement in the *Gazette* alleged that the Bradfords had "basely and villainously forged their own [almanac]." For their part, William and Andrew Bradford decried such "lying and abusive advertisements." Squabbling among rival philomaths and their printers was hardly unusual. By 1731, the Leeds almanac was once again sole property of the Bradfords. Their competitors were no longer a threat; Keimer and his successor each in turn went bankrupt and fled to Barbados, thus ending this clash of Titans.[9] But there was worse in store for Philadelphia's astrologer when he ran up against Poor Richard.

Leeds continued to include mental exercises in his almanac, unimpeded by criticism from the *Gazette*.[10] Showing no hard feelings, his almanac for 1732 poses a question credited to Thomas Godfrey: "The base of a plain triangle is 20, the perpendicular height is 7, and the angle opposite to the base is 80. 'Tis required to find the other sides and angles?"[11] Meanwhile Franklin had a falling out with Godfrey and thereafter decided to write his own almanac. The *Autobiography* calls Godfrey:

> a self-taught mathematician, great in his way, and afterwards inventor of what is now called Hadley's quadrant [a navigational instrument]. But he knew little out of his way, and was not a pleasing companion, as like most great mathematicians I have met with, he expected universal precision in everything said, or was forever denying or distinguishing upon trifles, to the disturbance of all conversation.

However, their differences ran deeper than this brief passage reveals. Before Franklin married, he courted a relative of Thomas's wife, who brokered the match. (The Godfreys were lodgers in Franklin's house.) A disagreement over the dowry severed their personal relations, and likely their business relationship as well, and soon afterward Poor Richard was born. Here, finally, begins the oft-told tale of Titan Leeds and his literary murder by Benjamin Franklin.

In 1732, there appeared *Poor Richard, 1733: An Almanack for the Year of Christ 1733*, by Richard Saunders, Philomath. The nom de plume derived from a combination of Poor Robin and Richard Saunders (variously, *Saunder*), whose own works are well represented in the Franklin collection of almanacs. Poor Richard Saunders would enter an already crowded market in Philadelphia, but was distinguished from the start by a superior wit. He begins by

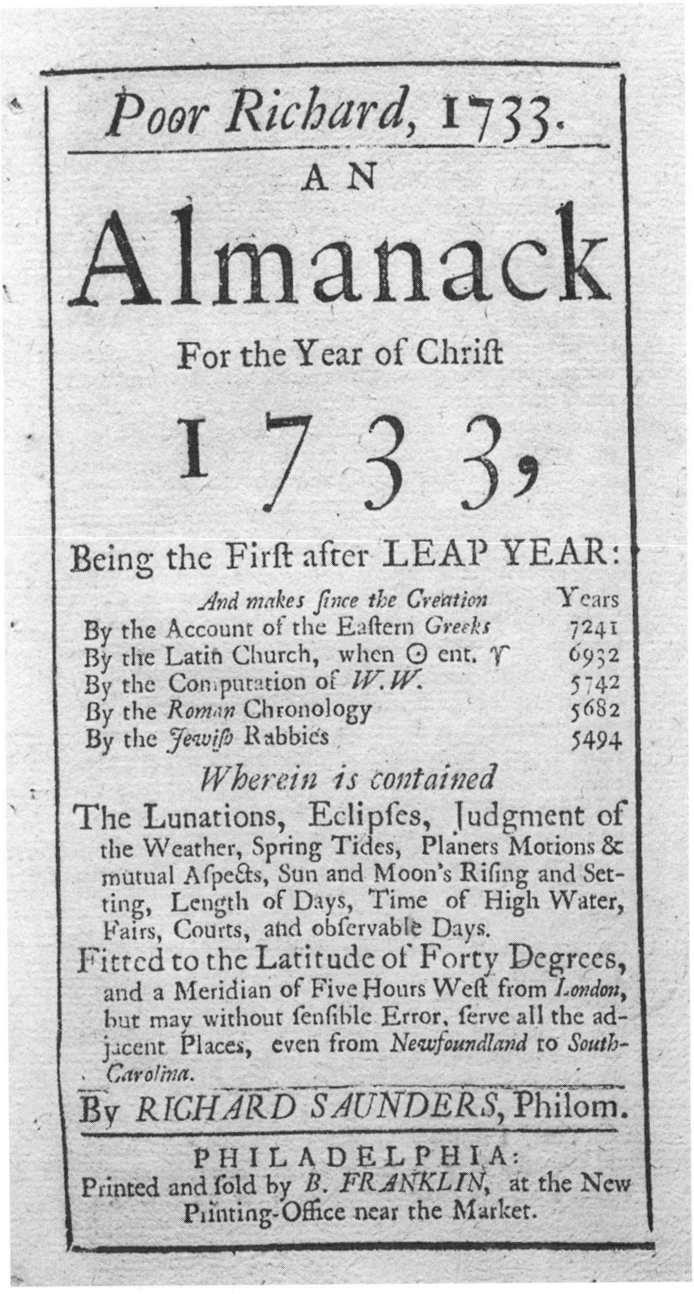

Poor Richard, 1733.

A N

Almanack

For the Year of Chrift

1733,

Being the Firft after LEAP YEAR:

And makes fince the Creation	Years
By the Account of the Eaftern *Greeks*	7241
By the Latin Church, when ☉ ent. ♈	6932
By the Computation of *W.W.*	5742
By the *Roman* Chronology	5682
By the *Jewifh* Rabbies	5494

Wherein is contained

The Lunations, Eclipfes, Judgment of the Weather, Spring Tides, Planets Motions & mutual Afpects, Sun and Moon's Rifing and Setting, Length of Days, Time of High Water, Fairs, Courts, and obfervable Days.

Fitted to the Latitude of Forty Degrees, and a Meridian of Five Hours Weft from *London,* but may without fenfible Error, ferve all the adjacent Places, even from *Newfoundland* to *South-Carolina.*

By *RICHARD SAUNDERS,* Philom.

PHILADELPHIA:
Printed and fold by *B. FRANKLIN,* at the New Printing-Office near the Market.

Fig. 3.2. Cover of the first *Poor Richard*. Rosenbach Museum & Library, Philadelphia.

admitting sheepishly that he intends to *profit* from this tract, but only to stop his poor but proud wife from destroying his books and mathematical instruments in frustration at his lazy stargazing ways. He would have published an almanac long before, but hesitated for fear that to do so would harm his friend and colleague Titan Leeds. This is no longer a consideration, however, as the stars portend that Leeds will soon perish, though the precise date and time is a matter of dispute. Poor Richard's prophecy specifies October 17 at 3:29 in the afternoon, "at the very instant of the ☌ of ☉ and ☿" (the conjunction of Sun and Mercury), while he claims that Leeds has forecast the dreaded event to occur nine days later.

One would expect that the controversy would have been resolved when Leeds' next almanac appeared, thereby confirming his nondemise and exposing the hoax while branding Poor Richard "a fool and a liar." Alas, Franklin would not be so easily swayed, for "there is the strongest probability that my dear friend is no more," and the newest pamphlet "may be only a contrivance of somebody or other." Later still, Poor Richard claimed to have received "much abuse from Titan Leeds deceased," and he contended that this was a case of a ghost who refuses to accept his own death.[12] Meanwhile, Richard's little book was well on its way to becoming the most successful American almanac of all time.

It's not that Franklin's work in this medium was wholly original. Far from it. Prior sources have been identified for most of his famous aphorisms and proverbs.[13] Many of these were improved by Poor Richard's editing. Some were revised for clarity, others for rhyme, all by Franklin's able pen. Thus "He is a greater liar than an epitaph" became "[He] can outflatter a dedication; and lie like ten epitaphs," while "Dine with little, sup with less, sleep high, and thou wilt live" was revised to "Dine with little, sup with less; Do better still, sleep supperless."[14] He updated the perplexing maxim "Let us lay aside fathers and grandfathers, and be good ourselves" to "Let our fathers and grandfathers be valued for *their* goodness, ourselves for our own," a more poetic sentiment which foreshadows Franklin's assertion that our personal accomplishments honor our ancestors but cannot be claimed by our descendants (chapter 1).[15]

Occasionally he corrects an absence of wit, as when "The clown, as well as his betters, practices what he censures, and censures what he practices," is revised to say, "Mankind are very odd creatures: One half censure what they practice, the other half practice what they censure; the rest always say and do what they ought."[16] The best proverbs have continued to evolve. For instance, although every American schoolchild learns that the almanac advised, "A penny saved is a penny earned," rather he wrote: "A penny saved is twopence clear. A pin a day is a groat a year."[17] (Or as my father used to say, "By the yard is hard, by the inch a cinch.") Closer to our popular revision of the proverb is the one found in Franklin's 1779 letter to an English merchant: "A penny saved is a penny got."

Not only the proverbs but the general format of Poor Richard's almanacs can be traced to predecessors like John Tipper, Tycho Wing, and the mathematically monikered Thomas Trigge. The same is true of the prefaces, poems, and other specific content: historical timelines, lists of kings, and mathematical items such as interest tables and currency conversion charts, all are inherited from the model almanacs in Franklin's personal collection. Even the Leeds hoax was not original, but was patterned after a similar deception by Jonathan Swift, author of *Gulliver's Travels*, who deflated a pompous philomath by nearly identical means.[18]

Like the British almanacs on which *Poor Richard* is modeled, each pamphlet includes the Golden Number and Epact. These are calendrically significant values that cycle with the years, and they may be calculated easily by anyone who has studied Sturmy's *Mariners Magazine*. In one installment, however, Franklin jokes that the Golden Number cannot be found, lamenting "I must content myself with a number of copper." Each year's calendar incorporates weather predictions and saints' days, and doubles as an ephemeris (describing the positions of various heavenly objects). In his second almanac, the calendar is followed by an apparent gibe at Messrs. Leeds and Godfrey:

Since the Eclipses take up so little space, I have room to comply with the new Fashion, and propose a *Mathematical Question* to the *Sons of Art*; which, perhaps, is not more difficult to solve, nor of less use when solved, than some of those that have been proposed by the ingenious Mr. G——y. It is this: A certain rich man had 100 orchards, in each orchard was 100 apple-trees, under each apple-tree was 100 hogsties, in each hogstie was 100 sows, and each sow had 100 pigs. Question, How many sow-pigs were there among them?[19]

The reference to his former friend Godfrey seems clear.[20] In case the italicized allusion in his preamble is too subtle, Poor Richard continues by copying Leeds's signoff, "Note: The answer to this question will not be accepted without its solution." At risk of overkill he adds, *Felix quem faciunt aliena pericula cautum,* or *Happy is he that learns to beware through the hazards of others.*[21] (*Felix* Leeds was credited with some of the counterfeit almanacs suffered by Titan. He was identified variously as father or brother to the more famous astrologer.[22])

The orchards-and-pigs challenge is easy enough. Franklin recognized that, for an almanac to have mass appeal, it had to be accessible to the mass audience. He could easily have taken a more difficult exercise from the mathematical books in the Library Company, but alienating his fellows was not sound business practice. One does not build America's most popular almanac by speaking gibberish. When Franklin drops a Latin phrase, it is one that is so well known as to require no explanation for the uneducated colonist.[23] The arithmetic problem is likewise not very challenging. Neither is it practical, in any direct sense; the rich man owns ten billion pigs, so the question is not meant to be of literal use. One simply multiplies repeatedly, and the answer follows. Such examples occur repeatedly throughout Franklin's almanacs. Some of these are quite similar to examples in Cocker's *Arithmetick*:

Profitable Observations and Notes: Three barley-corns make an inch, 12 inches a foot, 3 feet a yard, 5 yards and an half one pole or perch, 40 perches make a furlong, 8 furlongs make a mile, in a mile are 320 perches or poles, 1066 paces, 1408 ells, 1760 yards, 5280 feet; 63360 inches; 190080 barley-corns.[24]

Profitable Notes: In an acre are contained 4 square rood, 160 square pole, 4840 square yards, 43,560 square feet, 6,272,640 square inches. In a mile's length are contained 8 furlongs, 320 rod, 1760 yards, 5280 feet.[25]

In the machine at Derby in England for winding Italian silk, there are 26,586 wheels, 97,746 movements; 73,728 yards of silk wound every time the water-wheel goes round, which is three times every minute; 318,504,960 yards of silk in one day and night; and consequently 99,373,547,550 yards of silk in a year.[26]

These are simple examples, but then how much mathematical sophistication could be expected of the general reader at that time? (Or for that matter, in our own era? One is reminded of the modern publisher's advice to authors that each additional equation halves the readership.) Here is another bit of basic mathematics from Poor Richard, not using the multiplication principle alone, but of roughly equal difficulty:

> *A Frugal Thought.*
> In an acre of land are 43560 square feet.
> In 100 acres are 4356000 square feet;
> Twenty pounds will buy 100 acres of the proprietor,
> In £20 are 4800 pence; by which divide the
> number of feet in 100 acres; and you will find
> That one penny will buy 907 square feet; or
> A lot of 30 feet square.—*Save your pence.*[27]

Poor Richard proposed facetiously a gunpowder-driven sundial, guaranteed to serve "not only a man's own family, but all his neighbors for ten miles around." Carefully placed lenses would focus the sun's rays in order to set off one gun after the first hour, two guns to mark the second hour, and so on up to the twelfth hour. How many guns did he say would be required in all? (Can you think of a shortcut for adding up the first n counting numbers? This will be useful in chapter 8.)

Before Franklin wrote his pioneering demographic study *Observations Concerning the Increase of Mankind*, his alter ego Poor Richard foreshadowed those musings in the pages of the almanac. There we find extensive statistics on the population of the colonies, beginning with a table showing the results of the New Jersey census separated by age, gender, county, and slave or free status. (We northerners like to pretend smugly that slavery has always been a foreign concept here. That is patently untrue.) Summarizing, he lists the "Total of souls in 1737, 47369; ditto in 1745, 61403; increase 14034. *Query*, at this rate of increase, in what number of years will that province double its inhabitants?"

The New Jersey data is followed by a table enumerating new arrivals in the cemeteries of Philadelphia. Over a seven-year period, the city has seen 2,100 deaths, or on average 300 each year. (He carefully omits casualties of fatal shipboard illness among immigrants, which populate the cemeteries but do not properly belong in a calculation of the Philadelphia census.) Since the annual death rate is probably around 1 in 35, he can estimate the total population. It is not a flawless argument, but in the absence of truly reliable data, it is a clever and useful method. There is an appealing simplicity about his reasoning. (It's interesting to note that in 1999, the Supreme Court ruled against the use of an indirect method of enumeration in the United States Census, despite the advanced state of modern statistics and its omnipresent, unquestioned use in both industrial and scientific settings. Perhaps our most brilliant founding father and his unschooled readers are best left in their pre-Constitutional past.) His next example is equally simple and equally ingenious. Only incomplete data was available for Massachusetts, where the number of adult white men numbered 41,000. The total size of the populace was not recorded. However, in New Jersey, that same subgroup was known to comprise one-fourth of the total citizenry. Therefore, assuming the two colonies' populations to be distributed in roughly the same fashion, Franklin was able to approximate the size of his birthplace at 164,000 residents.

Poor Richard understood that population may increase exponentially (doubling at regular intervals), but whenever asking a mathematical question he simplified matters for the naive reader by assuming the growth to be merely "arithmetic" (adding the same number of people each year).[28] It's analogous to the difference between simple and compound interest: in the short term, one can safely confuse the two, but over time such an approximation produces gross inaccuracies. For example, he asks: "Let the expert calculator say, how long will it be, before by an increase of 64 *per annum*, 34,000 people double themselves?"[29] If indeed the increase is kept at 64 per year, it will take more than half a millennium. But if the increase is 64 in the first year, and growth continues to proceed at 0.1882% annually, the doubling will occur in under four centuries.

In the same almanac, we find some *combinatorics*, that branch of mathematics which counts patterns and arrangements of objects. Here Franklin remarks on the miracle of written language and the abundance of words:

> What an admirable invention is writing, by which a man may communicate his mind without opening his mouth, and at 1000 leagues distance, and even to future ages, only by the help of 22 letters, which may be joined 5852616738497664000 ways, and will express all things in a very narrow compass. 'Tis a pity this excellent art has not preserved the name and memory of its inventor.

The meaning of this passage is not immediately clear. First, you must understand that in mathematics, a "word" need not be meaningful, or even pronounceable. Rather, it is just a sequence of symbols taken from some prescribed set, an abstract notion that anticipates (as theoretical mathematics often does) unforeseen future applications such as genomes and computer password security. In the opinion of the great geometer N. I. Lobachevsky, "There is no branch of mathematics, however abstract, that will not eventually be applied to the phenomena of the real world."[30]

It took me some time to figure out what Poor Richard meant by the preceding example. Someone, likely an inattentive apprentice, must have miscopied one digit and outright deleted four others, including the one in front. For if we restore these:

_5852616738__497664000_
↑   ↑   ↑↑   ↑
2   0   88   0

then we get 25852016738884976640000.[31] That's precisely the number of ways to arrange twenty-*three* distinct letters so that no letter repeats and no letter is left out; the "22 letters" was a typo as well.[32] Most likely this idea was taken from Edmund Stone's *A New Mathematical Dictionary*, and the twenty-three letters may refer to the old Latin alphabet, before J, W, and U were fully independent symbols.[33]

Stone is explicitly cited as a source for another of Franklin's mathematical digressions, on the topic of perfect numbers. The latter concept goes all the way back to Euclid, possibly even to Pythagoras. The number 6 is called perfect because while its positive factors are 1, 2, 3, and 6, it is also true that $1 + 2 + 3 = 6$. The next smallest perfect number is 28, whose divisors are 1, 2, 4, 7, 14, and 28. (Note that $1 + 2 + 4 + 7 + 14 = 28$.) The third is 496. In two consecutive almanacs, Poor Richard challenges his audience to calculate more of them: "let the curious reader, fond of mathematical questions, find the fourth." While this represents a more challenging problem than counting sow-pigs, at least the question itself is easily understood. This one, it must be admitted, is too hard for the general reader, especially considering Franklin's spare exposition. However, the even perfect numbers are generated by a formula, $2^{p-1}(2^p - 1)$, which works so long as both p and $2^p - 1$ are prime numbers. Notice that $2^{2-1}(2^2 - 1) = 2(3) = 6$, $2^{3-1}(2^3 - 1) = 4(7) = 28$, and $2^{5-1}(2^5 - 1) = 16(31) = 496$, all of them perfect; while $2^{4-1}(2^4 - 1) = 8(15) = 120$, which is not perfect. (4 is not prime, and neither is $2^4 - 1 = 15$.) Stone mistakenly included 120 on his list, because he neglected to observe the prime number requirement. Franklin parroted the error, but to his credit, he printed a correction in his next almanac.[34] With the aid of our formula, you can generate as many even perfect numbers as you like. The odd perfect numbers are far more mysterious; even today it is not known whether there are any at all, but if so, they must be enormous.

Elsewhere, Poor Richard presents a related concept. The positive factors of 220 (aside from 220 itself) add up to 284. Meanwhile the positive factors of 284 (not including 284) sum to 220. For this reason 220 and 284 are called *amicable*. Franklin writes: "The second pair is 17296 and 18416. I shall be obliged to any of my readers that will tell me the third pair." It's not likely that any did so!

In the early years of the almanacs, an unsigned essay appeared in *The Pennsylvania Gazette* extolling the virtues of mathematics. First, arithmetic: ". . . 'tis well known that no business, commerce, trade or employment whatsoever, even from the merchant to the shopkeeper, etc. can be managed and carried on without the assistance of numbers; for by these the trader computes the value of all sorts of goods that he dealeth in, does his business with ease and certainty, and informs himself how matters stand at any time with respect to men, money, or merchandize, to profit and loss, whether he goes forward or backward, grows richer or poorer." Arithmetic "is useful for all sorts and degrees of men, from the highest to the lowest."

Meanwhile geometry is indispensable to astronomers, geographers, engineers, and others:

> 'Tis by the help of geometry, the ingenious mariner is instructed how to guide a ship through the vast ocean, from one part of the earth to another, the nearest and safest way, and in the shortest time. . . . By geometry, the surveyor is directed how to draw a map of any country, to divide his lands, and to lay down and plot any piece of ground, and thereby discover the area in acres, rods and perches. The gauger is instructed how to find the capacities or solid contents of all kinds of vessels, in barrels, gallons, bushels, etc. and the measurer is furnished with rules for finding the areas and contents of superficies and solids, and casting up all manner of workmanship. All these and many more useful arts, too many to be enumerated here, wholly depend upon the aforesaid sciences, viz. arithmetick and geometry.

The author points out that the methods of reasoning cultivated in these pursuits are widely applicable in areas completely unrelated to mathematics, and he admires ancient cultures which required such knowledge of their leaders, deeming all other societies "unfit to rule and govern." Whether or not the piece was written by Franklin, as editor and publisher he surely approved of its sentiments.[35]

There are no magic squares in Poor Richard's almanacs.[36] Yet it was shortly after the commencement of the almanacs that Franklin most likely began his great arithmetical experiments. In 1736, he became clerk of the Pennsylvania Assembly (see figure 3.3), a position he would retain until he was succeeded by his son William in 1751. (A second son, Francis, died in 1736 at the age of four.[37]) Benjamin's autobiography, drafted decades later, describes the tedium of his clerical position, which drove him to seek an elected office instead: ". . . I was at length tired with sitting there to hear debates in which as clerk I could take no part, and which were often so unentertaining, that I was induced to amuse myself with making magic squares, or circles, or any thing to avoid weariness." This must refer to some time early in his clerkship.[38]

Just what did those first arithmetic experiments consist of? An unlikely bit of oral folklore alleges that you can still view the original minutes of the House, their margins decorated with Franklin's scribbled squares. Alas, the reality is more mundane; the Pennsylvania State Archives confirms that the original notes were destroyed when those proceedings were published in 1754.[39] (The printer was Franklin and Hall. Can we surmise that Franklin supervised their destruction personally? No, because after 1747 Hall was the active partner.[40]) As expected, the printed minutes of the Assembly do show the interminable discussions endemic to all freely elected legislative bodies, but naturally Franklin's idle doodling has been expurgated from the record and thus from history.

Despite the disappearance of crucial evidence, nevertheless it is possible for us to reconstruct Franklin's early work. First he conquered the traditional style of magic square, wherein sums add equally across, down, and diagonally, but then he abandoned these "common and easy things" in favor of his own innovations.[41] These include many remarkable, even miraculous, examples. We will encounter his mature work later on in this book, but for now

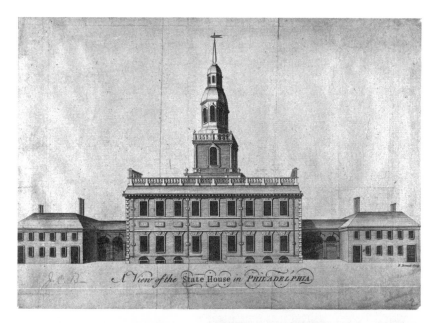

Fig. 3.3. The Pennsylvania State House, now better known as Independence Hall. Top: *A View of the State House in Philadelphia* engraved by R. Bennett, 1770, The Historical Society of Pennsylvania. Below left: The view from 2006. Below right: The Assembly Room, in which Franklin drew his magic squares and circles.

consider only the most basic "Franklin magic squares." The rows and columns of Franklin's matrix add up to the same number, as we would expect, but the two diagonals are replaced by four "bent rows," shown in figure 3.4.[42] (The term "bent row" was coined by Franklin.) In his words, "the diagonals are to be reckoned by halves,

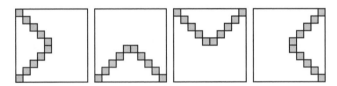

Fig. 3.4. Bent rows.

not crossing but turning at right angles from the center, by which 4 varieties are made instead of two."[43] To speak with mathematical precision, we should distinguish between the *semi-magic square*, with all rows and columns equal; the *fully magic square*, with rows, columns, and two diagonals equal; and the *Franklin magic* (or *bent row*) *square*, with rows, columns, and four bent rows equal. However, in general usage the term "magic square" may refer to any one of these types.[44]

Franklin's earliest mathematical experiments must have focused on the 4×4 case, where hundreds of bent row squares are possible.[45] One of these is fancifully portrayed in figure 3.5. (A complete list of all 4×4 Franklin-style magic squares is given at http://www.pasles.org/Franklin/4.html.) It takes only a moment to verify that the four bent rows, four rows, and four columns in figure 3.5 each sum to the same value, so this is a 4×4 Franklin magic square, albeit not necessarily one that was ever drawn by Franklin himself.

Some observations are in order. Notice where the values 1 and 16 fall in the completed matrix. These are the smallest and largest values, respectively, and they are located exactly two cells apart. The next-to-smallest value and the next-to-largest value are separated by the same distance. In fact all such "complementary pairs" in our magic square occur along a similar pattern. It can be proved that every 4×4 Franklin magic square obeys one of the pairings shown in figure 3.6, so long as we use the numbers 1, 2, . . ., 16.[46]

If you try your hand at building some 4×4 Franklin magic squares, you'll find the assembly process vastly simplified by using one of these five schematic alternatives. For example, if you guessed that the first row begins 1,2 then the rest of that row can only be composed of 15,16 or 16,15 (since the row sum must be

Fig. 3.5. *Square Deal II* (2006) imagines a bent-row square composed of samples of genuine American currency. (Franklin was contracted to print paper money for several of the original colonies.) Each row and column adds up to 34, as does each of the four bent rows. Such a matrix is called a *Franklin magic square.*

equal to 34). Only the first two diagrams in figure 3.6 allow for such a possibility, which narrows considerably the number of options one can pursue to fill in the rest of the square.

Look back at the Saturn, Jupiter, and Mars magic squares in chapter 2. Those are not Franklin magic squares, but their complementary

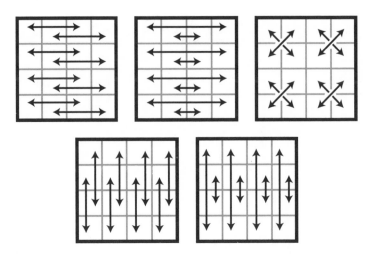

Fig. 3.6. Distribution of complementary pairs in a 4×4 Franklin magic square.

pairs are hardly scattered randomly throughout the matrix. Rather, they obey yet another blueprint: join each pair by a line segment, and its midpoint falls right at the central point of the square. Draw all of those line segments one on top of another, and you get an eight- or ten- or twelve-pointed asterisk! (Another way to say this is that the elements of any complementary pair are *symmetric with respect to the center.*)

Our simple example of a Franklin square pales before the magnificent baroque magic squares Franklin devised later on. He graduated to larger sizes where the 4×4 schematics no longer tell the whole story. Still, Franklin exploited an expanded version of the complementary "trick" in a far more subtle fashion when he created his great magic squares of size 8 and 16. We will confront those ultramagical squares and their complex properties in due time. Remarkably, Franklin's biographers completely miss the point when describing Franklin's magic squares. Time and again, they attribute to him the mundane straight-diagonal squares concocted by his predecessors, and neglect to notice the very innovation that makes his work so singular.

You may be wondering what really happened to old Titan Leeds, the Philadelphia stargazer. He continued to write almanacs for Franklin's competitor Andrew Bradford, until he finally died (*really!*) in 1738. Even so, the Bradfords continued to produce the "Dead Man's Almanac" in his name, proving that philomathic immortality is no illusion after all. (The considerate author had provided seven years' worth of advance calculations prior to his demise, or so his printer claimed.) As for Poor Richard, his almanac continued to much acclaim and profit until the late 1750s. But even he lived on— at first through an officially sanctioned almanac, in the hands of managing partner David Hall, and later through a host of unauthorized imitations. Even today you'll find Franklin's face on the cover of almanacs without even a remote connection to him. Old philomaths never die!

Notes

1. Titan Leeds, *The American Almanack For the Year of Christian Account 1731,* Printed and Sold by William Bradford in New York, and Andrew Bradford in Philadelphia, 1730.

2. I) 4.5 feet; II) 3 and 4 or vice versa; III) $20\sqrt{2}\sqrt[4]{3} \approx 37.2242$.

As evidenced by the published answers in his 1732 almanac, question II must assume for simplicity's sake that the earth is flat, which is why I have inserted a parenthetic clarification. Otherwise the values will vary depending on our initial latitude, which was not given.

Exercises are reprinted with the permission of NewsBank/Readex Inc. and the American Antiquarian Society.

3. The literal meaning of *philomath* is "lover of learning."

4. Another author whose almanacs were published by Franklin was John Jerman. (See chapter 8.) Jerman abandoned Franklin the same year as Godfrey, but he returned several times over the ensuing years.

5. *The Pennsylvania Gazette*, No. 114, Jan. 12–19, 1730–1731.

6. I suspect that these same questions were printed in Godfrey's almanac that year, but we may never know, because no copies survive. The answers to question 2 round to 0.2249 and 0.1418. Questions 1 and 3 are standard applications of spherical trigonometry.

7. Whitfield J. Bell, Jr., in his introduction to *The Pennsylvania Gazette 1728–1789*, Philadelphia: Microsurance, 1968.

8. Smyth, *Writings*, Vol. 1, p. 167.

9. To be more precise, it was a clash of *Leedses*; after Keimer stole the Titan name, Andrew Bradford attributed his product to Felix Leeds instead. (However, in 1729, an almanac identical to Bradford's "Felix" was published under the "Titan" name in Newport, Rhode Island, while Keimer's faux Titan continued in Philadelphia for one installment under the direction of David Harry. Therefore the joke stands.)

10. The influence of the *Ladies Diary* is unmistakable. Whereas the *Diary* classifies each exercise as a "mathematical question" or an "ænigma," Leeds simply replaces the latter category with "paradoxes." Solvers are acknowledged by name, as in the *Diary*. There are also differences. Leeds poses relatively few questions; his paradoxes are still mathematical, whereas the ænigmas are simply riddles; and only answers but not complete solutions are printed. Finally, his questions are less challenging.

11. The angles are 79.4773° and 20.5227°, the sides opposite them 19.9670 and 7.1197 units. Exercise reprinted with the permission of NewsBank/Readex Inc. and the American Antiquarian Society.

12. See the almanacs for 1734 and 1735.

13. The definitive references are Robert Newcomb, "The Sources of Benjamin Franklin's Sayings of Poor Richard," University of Maryland doctoral thesis, 1957; and Frances M. Barbour, *A Concordance to the Sayings in Franklin's POOR RICHARD*, Detroit: Gale Research Company, 1974.

14. Newcomb, pp. 57–58. This last proverb reflects the austere diet Franklin often prescribed, and actually practiced intermittently throughout his life.

15. Ibid., p. 107.

16. Ibid., p. 131.

17. *Poor Richard* for 1737. The saying continues: "Save and have. Every little makes a mickle." It is also found in a pseudonymously composed letter in the *Gazette* five years earlier. (The author was Ben Franklin, writing as "Celia Single.")

18. The victim was John Partridge, whose pamphlets are well represented in the Franklin collection, unsurprisingly enough. Partridge found that continued protestations were of no consequence. Like Franklin, Swift used a quasilogical argument to "prove" that the prophecy had been fulfilled, and both had the audacity to incorporate their victims' own accusing words in the process.

19. 10,000,000,000 pigs. That's more than ten times the number of pigs on the planet today! (Source: Food and Agriculture Organization of the United Nations, 2005.) Clearly this qualifies as a recreational question.

20. Again, it seems likely that Godfrey's early almanacs contained mathematical questions, but these have been lost to history. His later almanacs for 1733 and 1736 do not include any such challenges, but these two issues were written for another publisher and it may be that stylistic changes were imposed by the new printer.

21. The same proverb would be repeated in three more of Franklin's almanacs.

22. Wordplay was a favorite pastime of Franklin's; recall "Titan Pleiades" (previous chapter).

23. The expression in the previous paragraph appears in *The morall fabillis of Esope the Phrygian* (15th c.), Robert Wedderburn's *The Complaynt of Scotland* (16th c.), and Gailhard's *The Compleat Gentleman* (17th c.)

24. *Poor Richard* for 1745.

25. *Pocket Almanack* for 1745, also by Saunders/Franklin.

26. *Poor Richard Improved* for 1749.

27. *Poor Richard* for 1738. To be more precise, $4{,}356{,}000/4{,}800 = 907\frac{1}{2}$.

28. Clearly Franklin knew that populations need not grow arithmetically. For example, he assumes a constant doubling time of twenty or perhaps twenty-five years in his *Observations Concerning the Increase of Mankind and the Peopling of Countries* (see chapter 1), an assumption which corresponds to geometric growth (exponential increase). Lest there be any doubt, here are three more references from much later: *Papers*, 1973, Vol. 17, pp. 28–30 (esp. p. 29); 1983, Vol. 23, pp. 117–119; and 1984, Vol. 24, pp. 508–513.

29. *Poor Richard Improved* for 1750.

30. *Bulletin (New Series) of the American Mathematical Society*, Vol. 10, 1984, p. 151.

31. A more creative (but unnecessarily complex) explanation appears in Paul C. Pasles, "The Lost Squares of Dr. Franklin: Ben Franklin's Missing Squares and the Secret of the Magic Circle," *American Mathematical Monthly*, Vol. 108, 2001, pp. 489–511.

32. This is the multiplication principle at work. There are 23 possibilities for the first letter, 22 left for the next letter, then 21 choices, and so on. The answer is obtained by multiplying $23 \times 22 \times 21 \times \cdots \times 3 \times 2 \times 1$, commonly abbreviated 23! (read as "twenty-three *factorial*" and not actually shouted). In modern parlance, Poor Richard is counting the number of *permutations* of 23 letters.

33. It is a standard calculation, though, so another source is possible. For example, the same information can be found in Ozanam's *Recreations*. Also, *Divers Ouvrages* (see chapter 5) displays factorials up to $n = 22$; but the latter source was encountered too late to influence the 1750 almanac, if we accept that Franklin first saw this volume in 1750. A similar calculation appears much later in a book by James Ferguson, who would be the first author to publish Franklin's magic squares.

34. A less convincing explanation of the error is given in Pasles, 2001. Stone also states that "there are but ten [perfect numbers] between one, and one million millions," and Franklin repeats this misstatement. (The tenth perfect number has fifty-four digits, far outside the stated range.)

35. It is widely attributed to Franklin, and one particular line is still widely quoted as his: "What science then can there be, more noble, more excellent, more useful for men, more admirably high and demonstrative, than this of the mathematicks." Franklin's Masonic biographer suggests that Franklin read this essay at a meeting of St. John's Lodge. (It certainly carries Masonic overtones, with talk of "degrees of men," of Adam, and of Egypt.) However, the editors of the *Papers* (Vol. 2) and other experts (e.g., Lemay, in his *Documentary History*) deny that he was the author.

Franklin was still deeply involved with the *Gazette* in 1735, and so he must have approved of this essay if indeed he did not write it himself. Some of the details are similar to those in his *Proposals Relating to the Education of Youth* (1749) and the *Constitutions of the Publick Academy in the City of Philadelphia* (also 1749); see next chapter.

36. If I had a groat for every person who has claimed otherwise, I'd probably have around three shillings. But it is a simple matter to leaf through all twenty-six years of almanacs and verify that no magic square is contained therein.

37. The cause of death was smallpox. Whereas James Franklin's *Courant* had famously opposed smallpox inoculation, the adult Ben Franklin was a champion of this lifesaving if misunderstood medical practice. He took pains to contradict publicly the rumors that his son had been inoculated prior to contracting smallpox, and that this procedure led to his infection.

38. Around 1750 or 1751 he recalls having drawn magic squares in his "younger days," which leaves the timing open to wide interpretation. See chapter 2, endnote 56.

39. *Votes and Proceedings of the House of Representatives of the Province of Pennsylvania,* Philadelphia: B. Franklin and D. Hall, 1754.

40. From the *Autobiography*: "He took off all care of the Printing-Office, paying me punctually my share of the profits. This partnership continued eighteen years, successfully for us both."

41. The quotation is from a letter to Peter Collinson that we examine in greater detail in chapter 5.

42. A plethora of other properties are detailed in chapter 5. However, most of those properties are absent in a 6×6 magic square by Franklin that appears in chapter 8. Hence I will abide by the more minimalist definition given here. It describes the properties common to all of Franklin's known examples, with the exception of one that was apparently unoriginal anyway.

43. Letter of Franklin to John Canton, May 29, 1765.

44. The generic term refers to the semi-magic square in Harold M. Stark's *An Introduction to Number Theory,* to take just one example (Chicago: Markham Publishing Company, 1970, p. 118); to the Franklin magic square in Franklin's writings and those of his followers (chapter 9); and to the fully magic square (in most other sources).

45. Many of the reasons for believing that Franklin began by experimenting with the 4×4 case are detailed in my article "A Bent for Magic," *Mathematics Magazine,* Vol. 79, 2006, pp. 3–13.

46. *Ibid.*

*O*n occasion, American almanacs still presented mathematical challenges long after the era of Godfrey and Leeds. Many of the questions below are probably due to Nehemiah Strong, a Yale professor. The final four are from Abraham Weatherwise, a pseudonym for Philadelphia's own David Rittenhouse. A clockmaker by trade, the self-educated Rittenhouse was a gifted astronomer and instrument maker who built fantastic working models of the solar system. He succeeded Franklin as president of the American Philosophical Society. He was also a mathematician and served as first director of the United States Mint. [*Exercises are reprinted with the permission of NewsBank/Readex Inc. and the American Antiquarian Society.*]

• • • • •

A man had 10 sheep which he kept until they were 10 years old, they brought him an ewe lamb every year, and every [one] of those lambs, and their posterity, when one year old brought forth an ewe lamb. How many were the posterity of the 10 sheep, when 10 years old?

Four men owned 90 pounds between them, [so] that if to the first man's money you add z, it equals the second man's diminished by z, and the third man's multiplied by z, and the fourth man's divided by z, what was each man's part of the 90 pounds?

There are two numbers whose sum is equal to the difference of their squares; and if the sum of the squares of those two numbers be subtracted from the square of their sum, the remainder will be 60, what are the two numbers?

—Watson's Connecticut Almanack for 1777

• • • • •

A man filled his store with several sorts of grain, of which one half was wheat, one fourth rye, one eighth Indian corn, one tenth oats, and he had eight barrels of barley, how much had he of each and in the whole?

There are two numbers which are to each other as 5 and 6 and the sum of their squares is 2196. What are those numbers?

—Connecticut Almanack for 1781

• • • • •

How long time will a man be counting a million of millions at the rate of a hundred every minute?

What two numbers are those, the sum of whose sum and difference is 8, and 12 times the product of their squares is equal to that of their cubes?

A number is required; that the square shall be equal to twice its cube.

—The Farmer's Almanac for 1793

• • • • •

There are three numbers; the sum of their squares is 4992; the first being added to the third is 96, and the first added to the second is equal to the third; what are these numbers?

If a man lends out £100 at simple interest at 6 per cent per annum, and £1 at the same time and rate to receive compound interest—in what time will the amounts be equal? [Assume interest is compounded annually.]

A gentleman set out with a certain number of guineas in his pocket, but by accident lost 70 of them; but proceeding his journey he luckily found a purse of dollars, which contained just so many dollars as he had guineas when he set out: at the end of his journey he computed, that, if [subsequently] he had found half as many guineas as he lost, besides those dollars mentioned before, and [next] lost one quarter as many [guineas] as he found, he should then have just as many as he had when he set out; required his first number? [Assume an exchange rate of 4 dollars/guinea.]

Required the diameter of a vessel in form of a cylinder, which is 60 inches high, that will hold 100 wine gallons more than a square vessel of the same height, the sum of the sides of which is equal to the circumference of the cylinder? [Use 1 wine gallon = 231 cubic inches.]

—The New-Hampshire Calendar for 1795

Answers

1777

The number of sheep doubles every year, so the original ten have $10(2^{10}-1)$ progeny, not nearly as many animals as in Poor Richard's pig problem.

There are many possible solutions. The easiest approach if you want to keep the algebra really simple is to try a small value of z, like 1 or 2. For example the men might have 18, 22, 10, and 40 pounds, respectively.

6, 5 or -5, -6.

1781

$160 + 80 + 40 + 32 + 8 = 320$ barrels.

30, 36 or -30, -36.

1793

19,013 years.

4, 3 or 4, 0.

0 or $\frac{1}{2}$.

1795

40, 16, 56, or else 56, −16, 40.

114 years, 157 days. (Primers like *Hodder's Arithmetick* assume that a compounding period is one year, unless stated otherwise.)

175 guineas: $175 - 70 + \frac{1}{4}(175) + 35 - \frac{1}{4}(35) = 175$.

Approximately 47.79 inches.

4

Publisher, Theorist, Inventor, Innovator

Prodesse & delectare. E pluribus unum.
[Instruction & Pleasure. Out of many, one.]
— motto of the *Gentleman's Magazine* (collected
annual volume)[1]

$\mathcal{T}$he 1740s and 1750s encompass Franklin's period of most intense scientific creativity, specifically his inquiry into the nature of electricity. While others might rely on the patronage of monarchs and princes to support their research, Franklin obtained financial independence through his printing business, and so was able to retire from its active management and pursue his revolutionary experiments with greater zeal. Not every business venture, however, reflected the golden touch.

Following successes in the newspaper and almanac media, Franklin resolved to enter the nascent world of magazine publishing. As with the birth of the *Gazette*, this project had a shaky start. In 1740, Franklin described his proposal to a potential editor, who then took the idea to Andrew Bradford instead. (Bradford might have been sore over the loss of his postmaster's position to Franklin a few years prior, a designation of no little use to a printer.) Not to be deterred by the sudden competition, Franklin decided to serve as his own editor. The next year, both Franklin's *General Magazine* and Bradford's *American Magazine* debuted.

The first journal to use the term "magazine" was the *Gentleman's Magazine*, a British periodical that commenced publication in 1731. The word itself was borrowed from military jargon, where it denotes a storehouse of ammunition or gunpowder. The English *Magazine* collected items of popular interest culled from a diverse array of sources, hence its motto *E pluribus unum*. It contained proceedings of Parliament, geographic descriptions, poetry, "historical chronicles," and articles in natural philosophy. Franklin set out to produce an American incarnation of this tremendously popular journal, which had already inspired imitators in Britain proper. Imported books and magazines were sold by colonial printers alongside domestic publications, so the popular local appeal of such a magazine was already proven.

Franklin's editorial choices mirror the eclectic nature of the *Gentleman's Magazine* and other English journals. For the first six months of 1741, *The General Magazine, and Historical Chronicle, For all the British Plantations in America* carried speeches before Parliament and various colonial assemblies, plus other items of political interest, padded out by sundry articles including some original material. In the April issue, amidst that month's poetry selection we find "A Mathematical Question":

> A gentleman whilst walking in his ground,
> A stone, of shape and size uncommon found;
> Which having with a curious pleasure viewed
> A strong desire to measure it ensued:
> This by the judgment of an artist done,
> Its form was found to be an upright cone;
> The slaunting side of which was eighteen feet,
> The base diameter fourteen compleat.
> He strait have orders that it should be sent
> Into his garden for an ornament;
> Willing to have a room made in the same,
> In one of these three forms which he should name:
> A cube, a cylinder, or hemisphere,
> The greatest possible the stone can bear.

A skilful artist has proposed to cut
The same, at twenty pence the solid foot
The gentleman, all needless charge to save,
Some mathematical advice would have.
By which he may with satisfaction see,
Which has the least content of all the three;
That he may fix the form accordingly:
He likewise doth desire a just account
To what it will in sterling coin amount.[2]

(See figure 4.1.)

Unlike most items in the magazine, this one is not attributed to an earlier source. Perhaps the editor commissioned it from a local mathematician, or borrowed from elsewhere, or composed it himself. It is easy to imagine frugal Franklin (or parsimonious Poor Richard) asking us to minimize cost![3]

The May issue presented a solution, along with new problems proposed by the solvers. One of these was a navigation quandary of the type we have seen previously. Another query describes a game of chance between persons A and B, who alternately shake a box holding two dice. They have reached a point where the endgame would go as follows:

If A threw six before B threw seven, then A won the stake; but if B threw seven before A threw six, then B won it: Now, while A had the box, and consequently the next cast, an unlucky accident in the room obliged them to leave off play. The question therefore is, in what

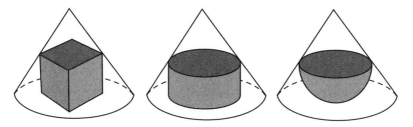

Fig. 4.1. Cube, cylinder, and hemisphere inscribed in a cone.

proportion the stake ought to be divided so that each man shall have his just share.

That question requires us to compare the probabilities that A (or B) will win. Can you solve this one?[4] Hint: You may use the formula for infinite series that was used by the Port Royalists (chapter 2): If $r < 1$, then $1 + r + r^2 + r^3 + r^4 + \cdots = 1/(1 - r)$.

The June issue included another solution, but no new problems, for this would be the final issue of the ill-fated *General Magazine*. Unlike the almanac and newspaper, this venture was the rare failure. If it was any consolation, his competitor fared worse, producing only three issues to Franklin's six.[5]

That same year saw the first publication, in England, of *The Gentleman's Diary, or the Mathematical Repository* (no connection with the similarly titled *Magazine*).[6] Like the *Ladies Diary*, it carried arithmetical, geometric, and trigonometric problems, together with clever "aenigmas" (riddles) of a nonmathematical stripe. Later issues added "paradoxes," rebuses, and other puzzles as well. This new *Diary* aimed to celebrate the "beauty and sweetness" of mathematics. Franklin collected every issue for the next eleven years.[7]

Questions from the *Gentleman's Diary*

Required the three least Numbers, which, if divided by 20, shall leave 19 for a Remainder; if divided by 19, shall leave 18; if divided by 18, shall leave 17; and so on (always leaving one less than the Divisor) to Unity?

• • • • •

WHAT's the greatest Solidity possible of that square Pyramid, whose slant Side is = 60 Inches? And the other Dimensions?

• • • • •

A Triangular Meadow I have to divide,
By three right Lines drawn from the midst of each Side;

Which Lines at a Pond in the Mead intersect,
Their Length is below*, from whence I expect
In Diary next Year, you'll to me be so kind,
As the Sides and Contents of this Meadow to Find.
* 15, 18, and 21 chains.

• • • • •

THERE is a regular Stone, whose Solidity = 288 Inches, and once the Length, with twice the Breadth, and thrice the Depth, is 36 Inches: Quære the Dimensions separate, and how many such Stones would wall Mr. Heath's Park; allowing its Area to be 97 Acres, and the Wall to be 7 Feet 4 Inches high, and 1 Foot thick. [Mr. Heath's Park is bounded by one line segment of length 3 furlongs and one circular arc. Only the arc is to be walled. See figure 4.2, left.]

• • • • •

There's a plain rectangled Triangular Field
Which one Acre and half, exactly doth yield
Its inscribed Square, hath likewise been found
Just ten Sixteenths of an Acre of Ground
From hence by a simple Equation produce
The Base, Cathetus and Hypotenuse. [Figure 4.2, center.]

• • • • •

[A circle is inscribed in a quarter-circle of radius 10, and then another circle is inscribed between them.] The Area of this little Circle is requir'd? [Figure 4.2, right.]

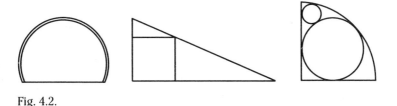

Fig. 4.2.

His business flourishing, and his career in public service now begun, Franklin had "abundant reason to be satisfied." Yet at this point he confesses two regrets: that the province possessed neither a militia for proper defense, nor a college for the education of young people. The first of these deficiencies was remedied by the raising of a militia, and through a battery of cannon funded by Franklin's lottery (in which he was assisted by James Logan, about whom we shall have more to say in the next chapter). The second concern would be resolved through his fortunate alliance with the Presbyterian Reverend Mr. George Whitefield.[8]

Whitefield, with his booming voice and careful enunciation, was reputed to have preached to twenty-five thousand people at a time, a claim which Franklin had doubted (along with "the ancient histories of generals haranguing whole armies"). A particular outdoor sermon offered the opportunity to determine the truth by experiment. While the minister spoke near the corner of Second and Market, Franklin backed through the crowd for almost a full city block, until he could no longer hear Whitefield's words. Assuming that one's listeners occupy a semicircle in front of the speaker, and that each person requires two square feet of standing room, he decided that "he might well be heard by more than thirty thousand." Contemporary maps show that the distance was around two-thirds of a furlong. Assuming that Franklin used a radius of one-half block to obtain his conservative estimate, can you check that his conclusion is correct? (*Note*: 1 furlong = 660 feet.) [9]

Franklin had published the reverend's sermons, reported on his ministry in the *Gazette*, and printed debates between Whitefield and his theological adversaries in the *Magazine*.[10] The minister's immense popularity failed to endear him to fellow clergy, who denied him a pulpit of his own. Thus his followers set out to construct their own house of worship and charity school, but financial woes prevented its completion and they were forced to sell it. Meanwhile Franklin's plan for a college had come to fruition, and he needed a building; the charity school was grafted onto the Academy of Philadelphia, which later became the University of Pennsylvania, the first "university" in America.[11]

Fig. 4.3. Revivalist at work. *George Whitefield* by John Greenwood
(1727–1792), after Nathaniel Hone, mezzotint on paper, 1769. National
Portrait Gallery, Smithsonian Institution.

The seeds of the idea can be found as early as 1743, though Franklin's final *Proposals* would not be published until six years later.[12] This and other founding documents reveal much of Franklin's educational philosophy. Naturally, his proposals cover a vast array of topics; the syllabus overwhelmingly favored secular knowledge, for in contrast to other American colleges of the time, its primary purpose was not to prepare the next generation of clergy. History is on the menu, so that the students can learn "the advantages of liberty and the mischiefs of licentiousness." And obviously mathematics plays an important role. He defers to John Locke on this.

> Mr. Locke is of [the opinion] that a Child should be early entered in Arithmetick, Geography, Chronology, History and Geometry. "Merchants Accounts, he says, if it is not necessary to help a Gentleman to get an Estate, yet there is nothing of more Use and Efficacy to make him preserve the Estate he has. Tis seldom observed that he who keeps an Account of his Income and Expences, and thereby has constantly under View the Course of his Domestic Affairs, lets them run to Ruin: And I doubt not but many a Man gets behind-hand Before he is aware, or runs farther on when he is once in, for want of this Care, or the Skill to do it. I would therefore advise all Gentlemen to learn perfectly Merchants Accounts; and not to think 'tis a Skill that belongs not to them, because it has received its Name [from], and has been chiefly practised by Men of Traffick [that is, merchants]."[13]

Compare the Constitutions of the Academy: ". . . As Matters of Erudition naturally flowing from the Languages, History, Geography, Chronology, Logick and Rhetorick, Writing, Arithmetick, Algebra, the several Branches of the Mathematicks, Natural and Mechanick Philosophy, Drawing in Perspective, and every other useful Part of Learning and Knowledge, shall be set up, maintained, and have continuance, in the City of Philadelphia. . . ."[14] Elsewhere Franklin outlines the plan for the Academy. "The hours of the day are to be divided and disposed in such a manner, as that some classes may be with the writing-master, improving their hands, others with the mathematical master, learning arithmetick, accompts, geography, use of the globes, drawing, mechanics, etc. while the rest are in the English school, under the English master's care."[15]

Franklin's *Proposals* points out that "Not only the *skill*, but the *habit* of keeping accounts, should be acquired by all, as being necessary to all." This is a theme to which he would return many times. To his daughter he wrote: "It will be of use to you if you get a *habit* of keeping exact accounts; and it will be some satisfaction to me to see them. . . . Study Poor Richard a little, and you may find some benefit from his instruction."[16] The almanac puts it quite succinctly: "Useful attainments in your minority will procure riches in maturity, of which writing and accounts are not the meanest."[17] In his guidelines as postmaster: "you are to make out a true and exact account thereof, with your disbursements, et cetera"; this might consist of "wages paid to riders, hire of horses," and the like.[18] A template includes detailed instructions to rival any modern tax form.[19] And then there is the tale of Franklin's partner in his South Carolina operations, a man both honest and educated, but a lousy bookkeeper:

> On his decease, the business was continued by his widow, who being born and bred in Holland, where as I have been informed the knowledge of accompts makes a part of female education, she not only sent me as clear a state as she could find of the transactions past, but continued to account with the greatest regularity and exactitude every quarter afterwards; and managed the business with such success that she not only brought up reputably a family of children, but at the expiration of the term was able to purchase of me the printing house and establish her son in it. I mention this affair chiefly for the sake of recommending that branch of education for our young females, as likely to be of more use to them and their children in case of widowhood than either music or dancing, by preserving them from losses by imposition of crafty men, and enabling them to continue perhaps a profitable mercantile house with established correspondence till a son is grown up fit to undertake and go on with it, to the lasting advantage and enriching of the family.[20]

She was not the only widow to find herself in charge of a press. The earliest female printers in America included both Cornelia

Bradford (Andrew's wife) and Franklin's sister-in-law. Ann Franklin carried on after James died on his thirty-eighth birthday, running the business alone until her son James Jr. returned from an apprenticeship under Uncle Benjamin; and she took over once again after her son's death. (An obituary shows that Ann survived all of her children.) Among the publications produced during their joint tenure was an almanac attributed to *Poor Job.*[21]

While James the younger apprenticed in Philadelphia, he and William Franklin (Ben's son) were tutored by Theophilus Grew, a mathematician of local renown who became the first mathematical professor at the Academy.[22] It has been suggested that Grew assisted in the preparation of astrological tables for Poor Richard's almanac.[23] His instructional services were advertised regularly in its pages, and he also wrote almanacs of his own. Some of these were authored under the pseudonym "Theophilus *Wreg*," possibly the least effective secret identity of all time.[24]

Careful bookkeeping was especially important in the printing business, where many transactions were conducted on credit. In addition, some debts were paid in barter, not currency, and this meant that a printer's store often stocked more than just books.[25] In addition, many of the books for sale were imported titles from English presses. Among the items advertised for Franklin's shop were slates, inkpowder, wax, account books, compasses, and protractors, and even a mariner's compass and other aids to navigation.[26] The books include *Ward's Young Mathematician's Guide*, Pardie's *Geometry*, and Love's *Surveying.* [27] A more extensive catalog of books sold by Franklin around this time lists several hundred volumes, some of which are mathematical in nature:

- Hatton's Arithmetick.
- Hill's Arithmetick.
- Hodgeson's Mathematics, 1st *Vol. Containing Geometry, Trigonometry, Navigation, Projection of the Sphere, Use of the Globes, And Logarithms.*
- Martin's Elements, of plain and lineal, spherical and conic Geometry, and Doctrine of Fluxions. *All demonstrated by Algebra, in a new and easy method.*

- Pardon's Arithmetick.
- Pemberton's View of Sir Isaac Newton's Philosophy.
- Ward's Posthumous Works, *containing his Navigation, practical and speculative Geometry, Surveying, Trigonometry and Doctrine of the Sphere.*
- Webster's Arithmetick.[28]

In 1743 the American Philosophical Society, a successor to the Junto, was established. Despite their personal differences, both Franklin and Thomas Godfrey were among the nine founding members. Godfrey was no longer authoring almanacs, but his services as a teacher of "navigation, astronomy, and other parts of the mathematicks" were advertised in Franklin's newspaper.[29] His was not an idle interest in navigation. Years earlier, around 1730, he had invented an improved mariner's quadrant to guide seafarers. Around the same time he contrived this device, an Englishman called John Hadley devised two quadrants, one of which was substantially similar to Godfrey's.[30] Comparable designs had been sketched—but never actually built—by Isaac Newton, Edmond Halley (of comet fame), and Robert Hooke.[31] Philadelphians cried foul when Hadley received the glory, but their appeals to the Royal Society were unsuccessful.[32] Decades later, for right or wrong, spiteful locals recalled this as a case of academic theft.[33] Godfrey died in 1749 at the age of forty-five of indeterminate causes; there are many references to his ill health, but few clues as to the nature of his physical ailments. (There is hearsay evidence that he was "apt to indulge in intemperate drinking."[34] In one letter Franklin himself writes, "I am inform'd [he] was so continually muddled with drink, that our Surveyor General . . . could never get him to assist in making the meridian line."[35]) The box on page 98 presents a particularly challenging calculus question Franklin placed in his *Gazette* in 1743, almost definitely written by Godfrey.[36]

Franklin seems to have been utterly convinced that the most subjective aspects of life could be measured, quantified, and thus made mathematical.[37] These are not limited to the social questions mentioned in chapter 1 (heredity, slavery, war, population growth, moral virtue), nor to the pamphlet in which he assumes that pleasure and

A Math Problem for Pirates,
from *The Pennsylvania Gazette*

Mr. Franklin,

As Privateering is now so much in Fashion, the printing [of] the following Question may be an Amusement, if not to the Privateers, yet to some of your Correspondents or Readers.

Suppose a Privateer, in the Latitude of 10 Degrees North, should at 6 in the Morning spy a Ship due South of her, distant 20 Miles; upon which she steers directly for her, and runs at the rate of 8 Miles an Hour. The Ship at the same time sees the Privateer, but not being much afraid of her, keeps on her Course due West, and sails at the Rate of 6 Miles an Hour; how many Hours will it be before the Privateer overtakes the Ship?

N.B. The Sailing is supposed on a Plane as plain Sailing, and the Privateer keeps her Course constantly directed toward the Ship.

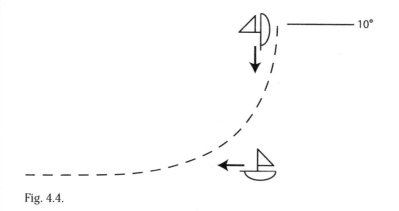

Fig. 4.4.

pain can be assigned units and compared using the law of transitivity (if $A = B$ and $B = C$, then $A = C$).[38] He brought quantitative reasoning to the realm of decision making at a time when this was still a radical notion, and he changed the way we measure time itself using a simple argument from mathematical economics to propose what would later be called "daylight saving time." The mathematization of life even extends to his humor. As "Alice Addertongue," a fictional *Gazette* correspondent, (s)he encouraged the publication of scandalous news:

> I mentioned above, that without good method I could not go through my business. In my father's lifetime I had some instruction in accompts, which I now apply with advantage to my own affairs. I keep a regular set of books, and can tell at an hour's warning how it stands between me and the world. In my daybook I enter every article of defamation as it is transacted; for scandals received in, I give credit; and when I pay them out again, I make the persons to whom they respectively relate *debtor*. In my journal, I add to each story by way of improvement, such probable circumstances as I think it will bear, and in my ledger the whole is regularly posted.[39]

Alice's love of scandal is rationalized arithmetically. Most likely, she claims, no more than one-fifth of a person's dirty laundry ever becomes public knowledge. Thus, if truth be told, she shows great restraint in exaggerating each item of gossip by only a factor of three, for this does not even begin to compensate for those indiscretions that were successfully concealed.

This sort of bookkeeping was meant only partly in jest. We have seen that Franklin kept a ledger of his own transgressions against a long list of virtues, ranging from temperance to humility. (A sample page was shown in chapter 1.) The latter quality must have been particularly difficult to maintain, for someone so widely accomplished; in the 1740s alone, he expanded his business, served in the militia, co-founded the American Philosophical Society, designed the Pennsylvania fireplace for indoor heating, and invented the lightning rod. In his memoir, he writes: "there is perhaps no one of our passions so hard to subdue as pride," and admits that "even if I could conceive that I had completely overcome it, I should probably be proud of my humility." Yet as Poor Richard opines, "A cypher [zero] and humility make the other figures and virtues of ten fold value."[40]

The instructions for this ledger of personal behavior were straightforward.

> I made a little book, in which I allotted a page for each of the virtues. I ruled each page with red ink, so as to have seven columns, one for each day of the week, marking each column with a letter for the day. I crossed these columns with thirteen red lines, marking the beginning of each line with the first letter of one of the virtues, on which line, and in its proper column, I might mark, by a little black spot, every fault I found upon examination to have been committed respecting that virtue upon that day.

Since Franklin's system was described in the most widely read autobiography in all of recorded history, it must have inspired countless others to copy his method of personal improvement. Many years later, the author Leo Tolstoy kept his own "Franklin journal" which contained a "table of weaknesses." (Instead of virtues, the headings listed vices to be avoided.) There he recorded "all [his] crimes, marking them with crosses in the corresponding columns."[41]

The table of virtues was not Franklin's only application of basic accounting principles to everyday life. Later in life, he also pioneered a decision-making technique that balanced the positive and negative consequences of each potential course of action. He referred to this as *moral algebra* or *prudential algebra*. It is not known precisely when he first devised this method, but late in life he claims to have "often practiced [it] in important & dubious concerns." The particulars are as follows. Make two columns corresponding to the reasons in favor of and against a particular option. Consider how much weight should be assigned to each reason, and then attempt to match those of equal weight. It may be that the first two items in the "pro" column are deemed of equal magnitude to the fifth item in the "con" column. Each set of reasons linked in this way—equal but of opposite value—may be crossed off and then disregarded.[42] When as many items as possible have been eliminated, the proper decision should be apparent. A simple example might look like this:

Should I enter into business with Mr. Smith?

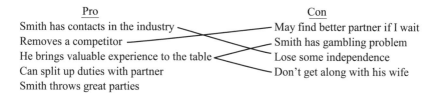

Pro	Con
Smith has contacts in the industry	May find better partner if I wait
Removes a competitor	Smith has gambling problem
He brings valuable experience to the table	Lose some independence
Can split up duties with partner	Don't get along with his wife
Smith throws great parties	

If we are fortunate enough to find one column empty, as here, the decision is clear: we should go into business with Mr. Smith. If neither column were empty, the decision should still be greatly simplified, and if we have sufficient confidence in our numerical assignments, we might simply total each column.

One description of Franklin's decision-making technique appears in a letter to grandnephew Jonathan Williams, who later became the first superintendent of the U.S. Military Academy at West Point. Franklin admonishes his younger relation, who is reluctant to settle down, advising him to practice the moral algebra: "if you do not learn it, I apprehend you will never be married." Williams replies that rather "I am afraid I shall never be married if I do, for the negative column seems in this instance the weightiest."[43]

In Franklin's calculations we find the roots of modern *utility theory*. He was far from the only such pioneer. Franklin's mentor James Logan promoted "social arithmetic," and before him Pascal and others.[44] Indeed, the notion of weighting an outcome by its value or desirability arises in many fields, among them philosophy, psychology, and economics. Desirability may be measured in terms of hedonistic pleasure, personal satisfaction, the glorification of God (as William Paley suggested), or the absence of pain.

The arithmetic underlying utility theory is basically the same as that used to calculate a weighted average or an expected value. For example, I am a university professor. At the start of the semester I may tell students that the midterm exam is worth 40% of their grade and the final exam 60%. If a student obtains scores of 85 and 95 points, respectively, then the student's semester average is (40% × 85) + (60% × 95) = 91 points, a weighted average.

Expected value is calculated in a similar way. Imagine that you are asked to play a dice-tossing game not unlike the one described in the *General Magazine*, earlier in this chapter. To simplify matters, let's say that you roll just once. If you obtain a 3, then you will receive twenty-one dollars. If you roll any other number, you will receive nothing. Assume that there is an entry fee of one dollar each time you play the game. The expected value of this game is $\left(\frac{1}{6} \times \$20\right) + \left(\frac{5}{6} \times -\$1\right) = -\$2.50$, which means that if you were to play repeatedly over a long period of time, you would expect to win an average of $-\$2.50$ per game. A similar argument can be used to explain why you should never play the lottery, which is sometimes described as "a tax on the mathematically illiterate."[45] (At the very least, one should view the ticket price as a charitable contribution.)

Arguing for the advantage of living a Christian life, Pascal used an analogous argument: If we believe in God and he does exist, our rewards are beyond imagining. If we do *not* believe and he *does* exist, the punishment is equally extreme. On the other hand, if we believe and are wrong, or we disbelieve and are right, there is little consequence either way.[46] Of course, one cannot simply choose to "believe," but Pascal thought that behaving as if we do will eventually lead to the real thing. Translated into the language of utility, we might assign values of $+40,000$ and $-50,000$ to eternal life and damnation, respectively. Living in accordance with Christian principles (that is, striving to believe) might be deemed an inconvenience, valued at -10, say. Living without regard for such principles could be termed neutral, assigned a value of zero. Even if the probability that God exists were as small as 1%, there is still a measurable advantage to belief:

$$\text{utility of belief} = 1\% \,(40,000) + 99\% \,(-10) = +390.1$$
$$\text{utility of disbelief} = 1\% \,(-50,000) + 99\% \,(0) = -500$$

Under our numerical assumptions, belief appears to be the favored strategy. Note, however, the difficulty of quantifying these inherently subjective concepts; perhaps some questions are beyond the reach of mathematical decision making. (You may find entertainment or even enlightenment by drawing up your own list of ingredients and repeating the calculation.)

A rational consumer seeks to maximize total utility. Just as important is the concept of *marginal* utility, the added satisfaction obtained by increasing consumption by one unit. Say you spend your leisure cash for the day on coffee and chocolate bars. Each successive cup of coffee brings you a little less additional satisfaction, and likewise for the chocolate, according to the law of diminishing marginal utility. Your last dollar spent on coffee should bring the same additional amount of satisfaction as the last dollar spent on chocolate, that is, their marginal utilities ought to be equal; this state is known as consumer equilibrium. Clearly, utility analysis plays a central role in modern economics; the typical textbook devotes an entire chapter to the subject.

It has been claimed that Jeremy Bentham, a founder of modern utility theory, was inspired by Franklin's moral algebra.[47] Bentham was personally acquainted with both of Franklin's scientific correspondents on the issue, Richard Price and Joseph Priestley, and he put forth his ideas shortly after Franklin communicated his own method to both men. Rather than simply maximizing one's personal utility, Bentham's *utilitarianism* sought to maximize the overall utility for the common good. (This guiding principle was reformulated and advanced by the political philosopher John Stuart Mill, son of Bentham's acolyte James Mill.) For instance, an environmental regulation that appears expensive today is intended to prevent thousands of catastrophic cancer deaths decades from now.[48] In the same way, when a doctor refuses to prescribe unnecessary antibiotics, she hopes to delay the evolution of drug-resistant pathogens. Both examples demonstrate a commitment to utilitarian goals.

Like his brief contributions to population science, Franklin's primitive ideas on utility brought quantitative methods to ingenious application; not by proving some sweeping theorem in higher mathematics, but by the extension of elementary arithmetic to a whole new world. It is brilliance in simplicity, applying arithmetic to seemingly unquantifiable situations. As with his more famous inventions, like the Pennsylvania fireplace, the lightning rod, and bifocals, he was simply ahead of his time. But the most significant contribution Franklin would make to the sciences was not in a primitive form of utility analysis, nor in the founding of what became America's first university, but in the entirely pre-mathematical realm of electricity.[49]

While visiting Boston, Franklin encountered a traveling lecturer named Spencer—misremembered later as "Dr. Spence"—who conjured various electrical tricks for colonial audiences.[50] Franklin observed a demonstration, and although he was impressed with the performance, he was convinced that it could be improved. Around the same time, by coincidence, the Library Company received an odd gift from their London agent: an "electrical tube," of which Franklin took possession.[51] (The tube accompanied a recent issue of the *Gentleman's Magazine* which carried an account of the latest electrical discoveries.[52]) Meanwhile the printing business continued to thrive, and between almanacs, newspapers, paper money contracts, and other ventures, Poor Richard had saved quite a few pennies (and groats and pounds). By 1748, he was successful enough to retire from active day-to-day management of his business and focus on his experiments alone. And so it was that a self-educated American, already well into middle age, became a scientist of international renown. The impact of this financial independence on Franklin, and thence the world, would be difficult to overestimate.

He could not have chosen a cleaner slate on which to begin. Whereas other areas of physics like mechanics and optics had enjoyed a long period of development, stretching from antiquity all the way to Isaac Newton, the study of electricity lagged far behind. (Its companion subject, magnetism, was nearly as stagnant; but the discovery of the lodestone and its maritime application led to the invention of various navigational instruments and some basic theory, so electricity was certainly the poorer relation.) The ancient Greeks had documented startling electrical properties of several substances. They knew that a piece of amber, when rubbed, would attract bits of silk and other light materials; this particular discovery has been attributed to Thales (circa 600 B.C.E.), traditionally known as the father of geometry and teacher of Pythagoras. Yet this phenomenon was not understood, even in any primitive way, and further discoveries were not forthcoming.

Remarkably, no significant additions to electrical knowledge were made for another two millennia, until the publication in England of the treatise *De Magnete* (1600). The author was William Gilbert, later court physician to Queen Elizabeth.[53] Gilbert described such concepts as electric force, charged body, and electric pole (analogous to the magnetic polarity already discovered).[54] He observed that many more substances are electrifiable, not only the amber and jet known to the ancients. Gilbert coined the very term *electric*, from ηλεκτρον ("electron"), the Greek word for amber.[55] He also invented an early electroscope.[56]

Over the next century, Gilbert's discoveries were extended by experimenters in England, Germany, France, and elsewhere. Bodies that may be electrified were classified as electrics. It appeared that others could not, and these were termed non-electrics. It was noted that electrics might not only attract, but also repel, a fact which was explicitly denied in Gilbert's treatise. Various electrical machines were invented. It was found that electricity flows, and could be transmitted (conducted) along a string or, better still, a length of wire. Eventually it became clear that all bodies might be electrified. Since both attraction and repulsion were possible, it was postulated that there must be two types of electricity; these were designated *vitreous* and *resinous*.[57] Armed with this knowledge, traveling lecturers offering electrical entertainments proliferated, among them Dr. Spencer, who encountered Franklin in Boston.

Nevertheless the discipline was still in its infancy when the Leyden jar was invented—by three different investigators working independently—in the mid-1740s. A jar coated with tinfoil, partially filled with water, with a metal rod inserted through the top and attached to the inner coating (see figure 4.5), it was capable of receiving and holding a powerful electric charge by some unknown mechanism. This was the state of affairs when our protagonist entered the scene, which brings us to what historian of science J. L. Heilbron calls "The Age of Franklin."

Fig. 4.5. Leyden jar.

Franklin's discoveries were wide ranging and profound. Not only his experiments, but his interpretation of them, changed the way scientists viewed electricity. Borrowing terminology from simple arithmetic, he identified *positive* and *negative* electricity, or sometimes *plus* and *minus*. This is not simply a change in terminology but an assertion that the vitreous and resinous types in actuality represent an excess or deficiency of a single electrical "fluid." As Franklin noted, it was not yet clear which sign really represented excess—we now know that the reverse would be a more accurate description—but his terminology stands today.[58]

As yet too primitive to be classified as a mathematical science, the study of electricity was on the threshold of its quantitative stage, and not only by virtue of those arithmetic signs. With his explanation of the Leyden jar, Franklin also appealed to the concept of *conservation of charge,* which states that the total charge in a system is constant: electricity is not created or destroyed, only transferred.[59] In Franklin's words, "the equality is never destroy'd, the fire only circulating."[60] The charges were not simply positive or negative, but assumed to be measurable amounts; Franklin repeatedly refers to quantities of "electrical fire" as increasing or decreasing. It was too early for the mathematization of this branch of physics, but just barely, as Franklin laid the groundwork for later developments. How could one hope to find a formula for capacitance until long after Franklin investigated the capacitor known as the Leyden jar? (Indeed, capacitance was once measured in units of "jars.") Likewise it has been argued that he foresaw the notion of electrical potential.[61] More immediately, an experiment suggested by Franklin and performed by one of his electrical followers, Joseph Priestley, led directly to the inverse square law.[62]

Priestley, a founder of modern chemistry, is often called the discoverer of oxygen. (One wonders what people were breathing for all those years prior to his discovery!) To him we owe the invention of carbonated water and the pencil eraser. His two-volume work *The History and Present State of Electricity* describes the aforementioned experiment, from which he concludes that "the attraction of electricity is subject to the same laws with that of gravitation, and is therefore according to the squares of the distances."[63] All other things being equal, if you double the distance, then the attraction is not halved, but rather quartered. Compare Coulomb's law $F = kq_1q_2/r^2$, where q_1 and q_2 are the respective charges, with Newton's law of universal gravitation $F = Gm_1m_2/r^2$, where charge is now replaced by mass. (G and k are appropriate constants, and r is the distance between the charged particles or objects.) In each, the distance appears in the denominator and is squared, which is why these are called inverse square laws.

All of this is interesting from a mathematician's point of view, but history would have to wait a little longer for the familiar mathematical laws that bear the names of Coulomb, Ampère, and Ohm. A true and accurate account of Franklin's role in the development of electrical science must focus on his experiments, particularly those which made him a worldwide celebrity.

While not the first to compare lightning with the electrical spark, Franklin described their similarities in greater detail and designed the first experiments to verify that these represent the same natural phenomenon.[64] Evidence for the analogy ranged from the mundane to the downright amusing. During one of his experiments, a chicken was electrocuted, then revived using artificial respiration (!) and found to have been blinded.[65] (Pigeons and a turkey were also victimized.[66]) The same effect had been observed among people struck by lightning, suggesting that the cause was also the same. A definitive answer required a more elaborate setup: A sentry box is placed atop a tower or steeple. A long vertical iron rod outside the box curls inside and attaches to a platform, on which the researcher (or his lucky confederate) stands during a thunderstorm. In this way a person may be electrified.[67] More famous is the kite experiment, which allows for a higher reach. Build a kite of wood and silk, with a wire extending from the top, and attach a key to the end of the kite string. Flown during a thunderstorm, the wet kite should draw electric fire from the clouds so that a jar may be charged from the key. This experiment was executed successfully by a number of researchers on both sides of the Atlantic, and Franklin's place in history was assured.[68]

It is impossible in this brief section to survey Franklin's electrical research satisfactorily. From him we derive the term *battery*, at least in its electrical sense. (Like *magazine*, the word has martial origins: the same public booster who raised funding for the battery of cannon for public defense, he also created a "battery" of lead plates and glass panes that he and his confederates charged electrically. This meaning evolved into the everyday definition which refers to a single cell, as opposed to a sequence of them.) He built electrostatic machines. He could detect a coming storm through a contraption rigged to his house, where atmospheric electricity caused bells to chime in warning. All of this and more was accomplished in a relatively short span of time, from his "retirement" till he was called to public service and left the realm of "philosophical amusements."

Of greatest immediate impact was his invention of the lightning rod, which saved countless buildings from destruction. This was no mere hypothetical threat. Year after year, homes and public buildings were lost to fire by lightning strike. Often the target was a church, since it was the highest building in many towns. (For some time, there were debates as to whether ringing church bells during a storm produced any protective effect. It was certainly not beneficial to the hapless bell-ringer.)

Before Franklin's letters on electricity were published in book form, they became known to the wider public through the pages of—you guessed it—the *Gentleman's Magazine*, which brings us back to where this chapter began. His writings quickly became the subject of debate in France, where the two-fluid nature of electricity was defended by Jean Antoine Nollet, an abbé. One of Franklin's most prominent early champions was the Comte de Buffon, who is better known to mathematicians for his famous "needle problem": If you drop a one-inch needle on a floor decorated with parallel lines one inch apart, what is the probability that the needle crosses a line? (See figure 4.6.) (The solution lies beyond the scope of this book, but the answer is entertaining: The probability is precisely $2/\pi$, or approximately 63.7%.[69]) Buffon arranged for a new French translation of Franklin's earliest electrical writings. His actions led directly to the confirmation of Franklin's conjectures by Dalibard and Delor in 1752.[70]

If Franklin's electrical inspirations were part of a grander theme in his philosophy, it is one of balance. Frequently he extols the value of *moderation*, which holds a place in the matrix of virtues. ("Avoid extreams," he advises in the *Autobiography*.) His moral algebra balances the pro and con, canceling until a path becomes clear. His *A Dissertation on Liberty and Necessity* claims that

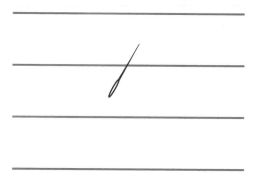

Fig. 4.6. Buffon's needle.

pleasure and pain occur in precisely equal amounts: both are quantifiable entities that ultimately cancel "when the accounts come to be adjusted," with God as the universal bookkeeper. This little pamphlet, written in 1725, presaged his thoughts on electricity, with pleasure and pain taking the place of + and −. Heilbron takes the analogy even further, corresponding *pleasure, pain,* and *insensitivity* to *positively charged, negatively charged,* and *grounded.* Amazingly, Franklin's thought experiment involving three "distinct beings" *A, B,* and *C,* in varying states of pleasure, pain and insensitivity, preceded his very real experiment (of three men exchanging charges) by two decades![71] The plus and minus charges are balanced, by virtue of the law of conservation of charge. When others urged rebellion from the King, Franklin prescribed loyalty, balance, and moderation for as long as he could, pursuing compromise and rapprochement until rebellion became inevitable. I think that we can even find this theme of balance in his magic squares. I refer not only to physical balance—if from each cell one suspends a tiny lead weight just as heavy as that cell number prescribes, this square plate will balance on its center—but also to a related property that we consider next.

I have argued elsewhere (http://www.pasles.com/Franklin.html) that the top two lines of Franklin's most famous magic square, in draft form, were originally written as

| 36 | 45 | 52 | 61 | 4 | 13 | 20 | 29 |
| 30 | 19 | 14 | 3 | 62 | 51 | 46 | 35 |

Notice that the pair of numbers in the middle of the top row, 61 and 4, add to 65. So does the next pair out in that row: $52 + 13 = 65$. In fact all such "complementary pairs" straddle the middle of the row in this way. That is true of the second row, too, and the rest of the magic square (not shown). This observation provides at least a partial explanation of the secret of this square's physical balancing.

Now decrease all of the entries by half of 65, which is the average value of all numbers in the matrix.[72] (You'll recognize this step if you completed the exercise at the end of chapter 2, where one can shift all entries by a constant value in order to solve the problem.) The magic is untouched, but watch what happens to the matrix:

3.5	12.5	19.5	28.5	−28.5	−19.5	−12.5	−3.5
−2.5	−13.5	−18.5	−29.5	29.5	18.5	13.5	2.5

Each complementary pair (like 61 and 4) has been transformed into opposites (like 28.5 and −28.5), and these lie an equal distance from the center of the row. Like equal but opposite charges, like pros and cons, like pleasure and pain, they balance perfectly.

And now it is time to step outside our roughly chronological narrative for a while and address Franklin's great magical accomplishments.

Notes

1. These words appeared at the front of each annual volume, which collected twelve monthly issues under a single binding. Both expressions are far older, but *E pluribus unum* was popularized in the eighteenth century by the *Magazine*. In 1776, when Franklin, John Adams and Thomas Jefferson designed the first draft of the Great Seal of the United States, they incorporated this same expression.

2. The cube has the least content, with a volume of around 238.91 feet. It costs approximately 19 *l*. 18 *s*. 2 *d*. (Recall from chapter 2 that in the 1700s one pound sterling was equal to 20 shillings, and each shilling equal to 12 pence.) We have assumed that each inscribed solid is oriented as shown in figure 4.1.

No calculus is required for this problem. Assuming that the cube rests on the base of the cone, only one inscribed cube is possible. That cube sits inside an inscribed cylinder, so clearly the *largest* inscribed cylinder is at least as large as the cube. It just remains to compare the cube and the hemisphere, which can be done using geometry alone, thereby avoiding any maximization techniques.

3. It is not reprinted from the *New England Weekly Journal*, which is the source of the preceding item in the magazine. Even if one accepts that Franklin lacked the geometric preparation to solve such a problem (and there is insufficient evidence to weigh in with any confidence on that issue), there was nothing to prevent him from inventing the question!

4. In any given toss of the dice, there are thirty-six possible outcomes. Five of these would give person *A* her desired sum of 6:

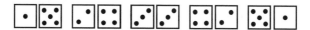

Thus the probability that A rolls a 6 on the first try is 5/36. The probability that A does not roll a six on the first try is 31/36. Thus the probability that B wins in this round is (31/36)(6/36), since six of the thirty-six possible outcomes yield a sum of 7. To complete the solution, you must consider the probabilities that A wins in one round, in two rounds, in three, and so on. Use $P(E)$ to denote the probability that the event E occurs. We already know that $P(A$ wins in the first round$) = 5/36$. Also $P(A$ wins in the second round$) = (31/36) \cdot (30/36) \cdot (5/36)$, because this would require that A loses the first toss, B loses the next toss, and then A wins the last toss. $P(A$ wins in the third round$) = (31/36) \cdot (30/36) \cdot (31/36) \cdot (30/36) \cdot (5/36)$, and so on. Therefore

$$
P(A \text{ wins}) = \frac{5}{36} + \left(\frac{31}{36} \cdot \frac{30}{36}\right)^1 \cdot \frac{5}{36} + \left(\frac{31}{36} \cdot \frac{30}{36}\right)^2 \cdot \frac{5}{36} + \cdots
$$

$$
= \frac{5}{36}\left\{1 + \frac{930}{1296} + \left(\frac{930}{1296}\right)^2 + \left(\frac{930}{1296}\right)^3 + \cdots\right\} = \frac{5}{36}\left\{1 \bigg/ \left(1 - \frac{930}{1296}\right)\right\} = \frac{30}{61},
$$

which is just over 49%. There is no need to go through such lengthy calculations to find $P(B$ wins$)$. Simply subtract: $P(B$ wins$) = 1 - P(A$ wins$) = 31/61$. The stake should be divided according to these proportions.

5. For its part, the *American Magazine* carried no mathematical items in its brief run, but this changed when a third generation of Bradford printers revived the title many years later. The new incarnation reprinted material from textbooks on geometry and trigonometry, and presented letters of Philadelphia mathematicians Thomas Godfrey and James Logan regarding Godfrey's invention of a navigational instrument. (At least half a dozen different magazines in prerevolutionary America shared the same title.)

6. To be honest, the apostrophe in the title did not appear until 1742.

7. Until 1752, that is, at which point he ceased collecting English almanacs of all titles. (Later in life he built a collection of French ephemerides.)

8. These sentiments are recounted in the *Autobiography*.

9. Half the block measures 1/3 furlong, or 220 feet. The area of a circle is $A = \pi r^2$, so the semicircle has area equal to $(1/2) \times \pi \times (220 \text{ ft})^2 \approx 76,000$ square feet, which will provide room for 38,000 listeners. The source is the *Autobiography*; see also the 1749 almanac.

10. *General Magazine*, No. 4, pp. 263–268; No. 5, pp. 313–317; No. 6, pp. 398–406.

11. University of Pennsylvania, University Archives and Records Center, http:// www.archives.upenn.edu.

12. *Autobiography*.

13. *Proposals Relating to the Education of Youth in Pensilvania*, Philadelphia: Franklin and Hall, 1749. Quotation marks notwithstanding, this is a slight paraphrase. The only change in meaning is at the beginning: "not necessary" was for Locke "not likely" (*Some Thoughts Concerning Education*, 1693).

14. *Constitutions of the Publick Academy In the City of Philadelphia*, 1749.

15. *Idea of the English School, Sketched Out for the Consideration of the Trustees of the Philadelphia Academy.* Franklin's sketch appears in Richard Peters, *A Sermon on Education . . .* , Philadelphia: B. Franklin and D. Hall, 1751.

16. Letter to Sarah Franklin Bache, April 6, 1773, *Papers*, 1976, Vol. 20, p. 141.

17. *Poor Richard Improved* for 1749.

18. *Papers*, 1962, Vol. 5, pp.161–168 and 1965, Vol. 8, pp. 412–413.

19. *Papers*, 1962, Vol. 5, pp. 169–177.

20. *Autobiography.*

21. Leona M. Hudak, *Early American Women Printers and Publishers, 1639–1820*, Metuchen, NJ and London: The Scarecrow Press, 1978.

22. Lemay, *Documentary History.*

23. For example, many library databases list Grew as a contributing author of *Poor Richard*. This claim is probably based on the testimony of Charles Evans.

24. *The Virginia Almanack, For the Year of our Lord God 1756.*

25. Hudak, cited in endnote 21.

26. These include quadrants and quarter waggoners, according to advertisements from 1734 and 1741.

27. "[Advertisement] Books sold by B. Franklin," *The Pennsylvania Gazette*, May 31, 1744.

28. *A Catalogue of Choice and Valuable Books*, 1744.

29. *Gazette*, October 30, 1740.

30. Frederick E. Brasch, "James Logan, a Colonial Mathematical Scholar, and the First Copy of Newton's *Principia* to Arrive in the Colony," *Proceedings of the American Philosophical Society*, Vol. 86, No. 1, Symposium on the Early History of Science and Learning in America, 1942, pp. 3–12.

31. Dava Sobel, *Longitude: The True Story of a Lone Genius Who Solved the Greatest Scientific Problem of His Time*, New York: Walker & Co., 1995, p. 90. See also J. Carson Brevoort's appendix to "Gerard Mercator: His Life and Works," *Journal of the American Geographical Society of New York*, Vol. 10, 1878, pp. 163–196.

32. Letters of Godfrey and James Logan to the Royal Society are reprinted in *The American Magazine and Monthly Chronicle for the British Colonies*, Philadelphia: William Bradford, Vol. 1, 1757–1758, p. 475. This printer was a grandson of the William Bradford whom we met in chapter 2. See endnote 5 of the present chapter.

33. Robert Patterson wrote to David Rittenhouse on April 18, 1794 of the quadrant "invented by our countryman, Mr. Godfry, but which has unjustly got the name of Hadley's quadrant" (*Transactions of the American Philosophical Society*, Vol. 4, 1799, p. 154). Earlier, James Logan alleged that Hadley actually stole Godfrey's idea, but it appears to be a case of simultaneous invention; if anything, it is likely that Hadley came upon his invention sooner, though this has not been proven. (Frederick B. Tolles, "Philadelphia's First Scientist: James Logan," *Isis*, Vol. 47, No. 1, 1956, pp. 20–30, esp. pp. 25–26.)

34. This is carefully labeled as hearsay evidence by Horace C. Richards, "Some Early American Physicists," *Proceedings of the American Philosophical Society* Vol. 86, No. 1, 1942, pp. 22–28.

35. Letter of Franklin to Cadwallader Colden, Feb. 13, 1750. *Papers*, Vol. 3, 1961, pp. 461–463.

36. *The Pennsylvania Gazette*, March 5, 1745. The question is signed T. G., which may denote Thomas Godfrey or Theophilus Grew, most likely the former. (An earlier letter to the *Gazette* [August 15, 1734] signed "T. G." is attributed to Godfrey, and a letter of James Logan to Franklin [May 1, 1731] refers to a "T. G." who is almost certainly the same person.)

37. Perhaps some would object that basic arithmetic does not warrant the name "mathematics," to paraphrase Truman Capote's complaint "That's not writing, that's typing." However, the use of basic arithmetic as applied to demography and decision making, to take two examples, surely was a necessary first step before anyone would bring more sophisticated mathematical tools to the table in those particular pursuits.

38. Benjamin Franklin, *A Dissertation on Liberty and Necessity, Pleasure and Pain*, London, 1725. A precise objective measure of pain still eludes medical science today.

39. *The Pennsylvania Gazette,* September 12, 1732. Smyth and others attribute the letter to Franklin.

40. *Poor Richard Improved* for 1750.

41. Eufrosina Dviochenko-Markov, "Benjamin Franklin and Leo Tolstoy," *Proceedings of the American Philosophical Society,* Vol. 96, No. 2, 1952, pp. 119–128. The extent of Franklin's influence on Tolstoy is the subject of continuing debate.

42. Franklin to Jonathan Williams, Jr., April 8, 1779, *Writings*, Vol. 7, pp. 281–283.

43. Prev. cit. Also Williams to Franklin, April 13, 1779, *Papers*, Vol. 29, 1992, pp. 318–319.

44. Tolles, 1956, pp. 21–22. Daniel Bernoulli (an early probabilist) and Adam Smith had similar ideas. Coincidentally, one of the most significant early utilitarians was Claude-Adrien Helvetius, whose widow befriended Franklin in his later years but rejected his proposal of marriage.

45. Actually it's called a "tax on the stupid," but that is not quite accurate. Ignorance should never be confused with stupidity.

46. Jonathan Baron, *Thinking and Deciding*, Cambridge: Cambridge University Press, 1988.

47. Jacob Viner, "Bentham and J. S. Mill: The Utilitarian Background," *The American Economic Review*, Vol. 39, No. 2, 1949, pp. 360–382.

48. I am reminded of an evening news item, aired across the country at the end of 2005. The gist was that private industry was much more efficient than the federal government at post-hurricane rebuilding efforts, partly because the latter required asbestos inspectors to review each building. However, a simple utility calculation would suggest that such inspections are ultimately the cheaper option, given the high financial and personal cost of deaths due

to mesothelioma. (Ironically, a later segment on the same program was devoted to a vitamin that "may" prevent cancer; yet the link between asbestos inspections and cancer prevention is certainly far better established.)

49. Penn was the first college in America to declare itself a university. It was not, however, the first college.

50. Franklin's autobiography calls him "Dr. Spence," but I. Bernard Cohen identifies him as "Spencer" in *Benjamin Franklin's Science*. Cohen also questions the date of their meeting, which Franklin places in 1746.

51. In *Benjamin Franklin's Science*, Cohen suggests that Franklin has reversed the sequence of events, that the Library Company's electrical tube arrived before Franklin saw Spencer's demonstration, and that these events might even have been separated by several years.

52. J. L. Heilbron, *Electricity in the 17th & 18th Centuries: A Study of Early Modern Physics*, Berkeley–Los Angeles–London: University of California Press, 1979, p. 325.

53. Herbert W. Meyer, *A History of Electricity and Magnetism*, Cambridge, Mass.: The MIT Press, 1971.

54. A comparison of electric and magnetic attraction was made by Girolamo Cardano just a few decades before. See Heilbron, p. 174.

55. Sir Edmund Whittaker, *A History of the Theories of Aether and Electricity, I: The Classical Theories. The History of Modern Physics, 1800–1950*, Vol. 7, New York: American Institute of Physics, 1987 (reissue of Whittaker's second revised edition, 1951).

56. Brother Potamian and James J. Walsh, *Makers of Electricity*, New York: Fordham University Press, 1909.

57. Whittaker, *A History of the Theories of Aether and Electricity I*.

58. William Watson (in England) also came up with the notion of "plus" and "minus" electricity, but he lacked a theoretical justification for this hypothesis. See Heilbron, p. 329.

59. Heilbron (p. 330) debunks the frequent claim that Franklin and his coexperimenters in Philadelphia actually *discovered* the fact that charge is conserved. However, he was the first to make use of this concept.

60. Franklin to Peter Collinson, May 25, 1747, Letter II in *Experiments and observations on electricity, made at Philadelphia in America, by Benjamin Franklin, L.L.D. and F.R.S.; to which are added, letters and papers . . .* , London: Printed for David Henry . . . , 1769 [4th edition].

61. Heilbron, p. 329, footnote 14, citing Mario Gliozzi, *L'elettrologia fino al Volta*, Vol. 1, 1937, pp. 188–189.

62. Robert E. Schofield, introduction to Joseph Priestley, *The History and Present State of Electricity [The Sources of Science, No. 18]*, Vol. 1, New York and London: Johnson Reprint Corporation, 1966, p. xxxvi; also Vol. 2, pp. 372–376.

63. Priestley, Vol. 2, p. 374.

64. A list of similarities Franklin composed in 1749 is recounted in his 1755 letter to John Lining, in *Writings*, Vol. 3, p. 255.

65. More specifically, he refers to a *pullet*, which is a young hen.

66. It took five Leyden jars to execute a turkey, which was then found to be delicious ("uncommonly tender").

67. *Opinions and Conjectures, concerning the Properties and Effects of the Electrical Matter, arising from Experiments and Observations, made at Philadelphia,* 1749. *Writings*, Vol. 2, pp. 427–454.

68. Whether Franklin himself ever performed the kite experiment has been a matter of debate. See Cohen, *Benjamin Franklin's Science* and Tom Tucker, *Bolt of Fate: Benjamin Franklin and His Electric Kite Hoax* (New York: Public Affairs, 2003) for two sides of the issue.

69. The problem is usually phrased in a more general fashion, so that the needle's length can vary in relation to the distance between lines. See Stewart, p. 621, No. 11–2.

70. Heilbron, pp. 347–348.

71. Heilbron, p. 329, footnote 14.

72. Compare the theorem on p. 5 of Pasles, "A Bent for Magic."

5 A Visit to the Country

For want of a nail the shoe was lost; for want of a shoe
the horse was lost; and for want of a horse the rider was
lost. . . .
—*Poor Richard Improved* for 1758

*T*he best-known mathematical episode in the life of
Benjamin Franklin involves a visit to the home of his friend and colleague James Logan. Though portrayed in the mathematics literature as simply a personal acquaintance of no particular consequence, Logan is better known to local historians as one of the most
important political and scientific personages in early America.

Having emigrated from Bristol in 1699 as secretary to William
Penn, Logan would go on to hold a variety of positions at the very
highest levels of the provincial government. The year Franklin became Assembly clerk, for example, Logan was simultaneously acting governor of the colony and chief justice of its supreme court.
After sufficient success in business, he was able to pursue science
with a passion unmatched among his compatriots. In support of his
scholarly activities, he assembled what was surely the best private
library in America at the time, and he is alleged to have owned the
first copy of Newton's *Principia* in the New World.[1] At a time when
many libraries were largely focused on theology, his was scientifi-
cally well-seeded. Mathematics, astronomy, optics and botany were

among Logan's scientific interests, and it was in the last of these that he made the most noteworthy achievement. He was Franklin before Franklin: businessman, public servant, scholar, and the most successful American bibliophile of his era.[2]

Logan inspired and advised young scientists such as Franklin, John Bartram, Cadwallader Colden, and Thomas Godfrey in matters of natural philosophy, before each of them made their individual contributions to botany, physics, and mathematics. Without the encouragement of their distinguished mentor, any one of these talents might have gone undeveloped and unrealized. Every bit as important as Logan's advice were his contacts in the Old World, most notably the English botanist and electrician Peter Collinson and the Swedish naturalist Carolus Linnaeus, who initiated the Latin naming system by virtue of which we are *homo sapiens* and our dogs are *canis familiaris*.[3] For their botanical observations, Logan and his circle were well-rewarded, as many Linnaean appellations today have a familiar ring: *logania, bartramia, collinsonia, franklinia* (figure 5.1).

Fig. 5.1. *Franklinia alatamaha* and *Collinsonia canadensis*, American Philosophical Society Library. The latter plant is still used in popular folk remedies today. Both sketches are from the hand of second-generation naturalist William Bartram.

The family *Loganiaceae* consists of dozens of genera comprising hundreds of species, including ornamental and medicinal plants, among them the sources of strychnine and the arrow poison curare.

This coterie of descriptive botanists traded seeds back and forth between the continents, but it is nothing so mundane as mere classification that secures Logan's place in the history of biology. In a more theoretical vein, his experiments with American corn provided confirmation of the new hypothesis describing the sexual reproduction of plants.

Logan was a correspondent of Edmond Halley and John Flamsteed, the first and second Astronomers Royal, whose works he studied. His letters, and the annotations in his books, show a mind well versed in the sciences. His skill in operating mathematical instruments to map boundary lines was indispensable in his public and private roles as Penn's agent and as a buyer and seller of properties. Like Franklin, he attempted to apply abstract mathematical thinking to moral issues.[4] For his work in so many fields, he has been called "Philadelphia's first scientist," "one of those who made the Quaker City the scientific capital of colonial America," and "the first mathematical scholar in the American colonies," while the *Dictionary of Scientific Biography* argues that he was "the most intellectually capable scientist of colonial America."[5]

As a mentor in scientific and business affairs, as a model of the many-faceted scholar, and as a conduit to prized contacts in the Old World, James Logan was a pivotal figure in Franklin's life.[6] Without their friendship, perhaps Franklin would not have had the opportunity to so impress Peter Collinson, which could have led to momentous changes in the historical timeline. (Franklin knew Collinson as the London agent for the Library Company, but whether he would have made the same impression without Logan's encouragement of his scientific side or Logan's frequent cheerleading on his behalf is impossible to say.) Collinson brought Franklin's discoveries to the attention of the Royal Society, proposed his election as a Fellow, and was responsible for the publication of his *Experiments and Observations on Electricity*, sealing Franklin's reputation as the

Fig. 5.2. Above: *Peter Collinson,* by J. S. Miller, Burndy Library, Dibner Institute for the History of Science and Technology. Note botanical specimens and book on display. Facing page: James Logan, courtesy of The National Society of Colonial Dames of America in the Commonwealth of Pennsylvania at Stenton, Philadelphia.

preeminent electrical scientist of his day. To see the crucial role played by Collinson in this tale, simply read these words Franklin wrote to him in 1747:

Sir,

Your kind present of an electric tube, with directions for using it, has put several of us on making electrical experiments, in which we

Fig. 5.2. (continued)

have observed some particular phænomena that we look upon to be new. I shall, therefore communicate them to you in my next, though possibly they may not be new to you, as among the numbers daily employed in those experiments on your side the water, tis probable some one or other has hit on the same observations. For my own part, I never was before engaged in any study that so totally engrossed my attention and my time as this has lately done; for what with making experiments when I can be alone, and repeating them to my Friends and Acquaintance, who, from the novelty of the thing, come continually in crouds to see them, I have, during some months past, had little leisure for any thing else.[7]

Franklin's scientific reputation preceded him in France, where he was sent to plead the colonies' case in 1776. It is often said that without having already distinguished himself as a scientist of international stature, he would not have been so warmly received in France, nor as successful in negotiations there.[8] Certainly others in the diplomatic mission who lacked such credentials, notably Arthur Lee and John Adams, experienced far greater difficulties.

Consider, then, this delicate chain of reasoning: without Logan, Franklin might not have realized his potential as a scientist, or could have done so without making an impact; and without his scientific reputation to stand on, this diplomat might not have been welcomed into French society. When he arrived on the Continent in 1776, his reputation as a scientist was well established, and this standing aided in his efforts to negotiate treaties and acquire loans for the American cause. Absent the alliance with France, the nascent nation might well have been stillborn. By that logic, then, James Logan was an unwitting hero of the Revolution, though he died decades before the first shots were fired. Those who believe that history can pivot on the actions of just one person may find this argument more plausible than those who see the tides of history only on the largest scale, riding inexorable and unyielding forces toward a preordained outcome. But those skeptics who say that history does not subtly turn on single events or the actions of great personalities would do well to recall that . . . *for want of a rider the battle was lost, for want of that battle the war was lost, for want of the war the kingdom was lost, and all for the want of a horseshoe nail.*

In any case, the dominoes did fall, for Franklin did befriend James Logan, and thus it was that—soon after Logan's death in 1751—Franklin would recount one of their last meetings in a letter to Collinson. The setting was Stenton, Logan's Germantown estate (figure 5.3).

That Georgian-style home still stands, the 500-acre estate now dwindled to three. But it survives only by a miracle. For while Franklin's lightning rod may afford protection from the fire in the sky, we remain helpless before the infernos of that human evil

Fig. 5.3. Scenes from Stenton, the setting for the famous magic square meeting between Franklin and Logan. Much of the library was likely housed in the Blue Lodging Room, lower right. Interior photos appear courtesy of The National Society of Colonial Dames of America in the Commonwealth of Pennsylvania at Stenton, Philadelphia.

called War. In 1777, as Mars blazed in Philadelphia, two British soldiers came to Stenton under orders to burn it to the ground. Other area mansions would fall prey to that fate, but Stenton was spared. According to oral tradition, the reason was the intervention of the quick-thinking housekeeper, a former slave of the Logan family who had stayed on as a paid servant. She quickly reported that two deserters were hiding at Stenton, and the soldiers were arrested by their own deceived comrades before the arson order could be executed.[9]

But all of this was in the distant future when Franklin trekked to Stenton one day, as he recounts in a letter to Collinson. Many years later, this missive was published in the fourth edition of *Experiments and Observations*, as reproduced below:

SIR,

ACCORDING to your request, I now send you the Arithmetical Curiosity, of which this is the history.

Being one day in the country,[10] at the house of our common friend, the late learned Mr. *Logan*, he shewed me a folio *French* book, filled with magic squares, wrote, if I forget not, by one M. *Frenicle*, in which he said the author had discovered great ingenuity and dexterity in the management of numbers; and, though several other foreigners had distinguished themselves in the same way, he did not recollect that any one *Englishman* had done any thing of the kind remarkable.[11] [See figure 5.4.][12]

Logan's magnificent library was uncharacteristically strong in mathematics and the sciences, and thus it is not surprising that Logan had acquired a copy of Frénicle's work. (Bernard Frénicle de Bessy's investigations of the magic square appeared posthumously, together with work of other authors, in *Divers ouvrages de mathematique et de physique*, 1693.)[13]

So, had the English conceived anything so remarkable? Franklin does not allow this jab at their fellow countrymen to go unanswered.

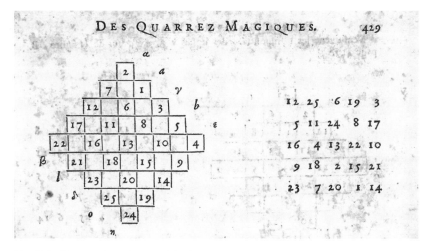

Fig. 5.4. Excerpt from Frénicle's *Des quarrez ou tables magiques.* Taken from the actual copy perused by Franklin. The Library Company of Philadelphia.

This accomplished magic-square-maker momentarily assumes another role, the exponent of practicality:

> I said, it was, perhaps, a mark of the good sense of our *English* mathematicians, that they would not spend their time in things that were mere *difficiles nugæ*, incapable of any useful application.[14] He answered, that many of the arithmetical or mathematical questions, publickly proposed and answered in *England*, were equally trifling and useless.

Here we enter the acrimonious debate that continues today: what constitutes worthwhile knowledge? Why learn anything that is not immediately applicable to the real world? Or as the students say, "When am I ever going to use this?" His resolution could not be more precisely put by any modern educator:

> Perhaps the considering and answering of such questions, I replied, may not be altogether useless, if it produces by

practice an habitual readiness and exactness in mathematical disquisitions, which readiness may, on many occasions, be of real use. In the same way, says he, may the making of these squares be of use.

Having accepted that this recreational activity may serve a concrete purpose after all, Franklin now admits that the topic is of some personal interest.

I then confessed to him, that in my younger days, having some leisure, (which I still think I might have employed more usefully) I had amused myself in making these kind of magic squares, and, at length, had acquired such a knack at it, that I could fill the cells of any magic square, of reasonable size, as fast as I could write them, disposed in such a manner, as that the sums of every row, horizontal, perpendicular, or diagonal, should be equal; but not being satisfied with these, which I looked on as common and easy things, I had imposed on myself more difficult tasks, and succeeded in making other magic squares, with a variety of properties, and much more curious. He then shewed me several in the same book, of an uncommon and more curious kind; but as I thought none of them equal to some I remembered to have made, he desired me to let him see them; and accordingly, the next time I visited him, I carried him a square of 8, which I found among my old papers, and which I will now give you, with an account of its properties. [See figure 5.5.]

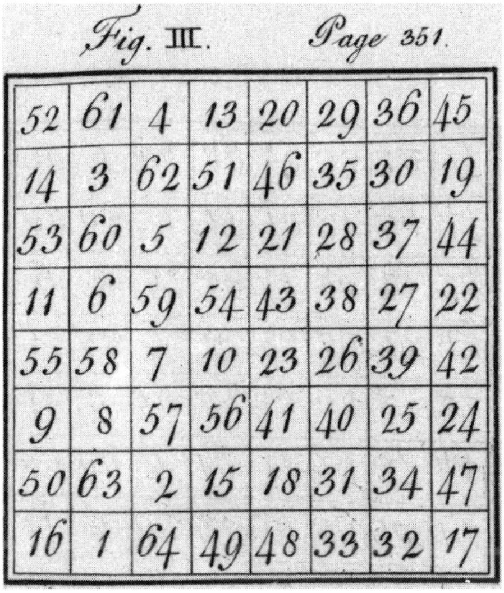

Fig. 5.5. ". . . a square of 8, which I found among my old papers." Detail from *Plan of Whirlwind,* Plate II, p. 226, in Franklin's *Experiments and Observations on Electricity,* 1769, The Historical Society of Pennsylvania.

The properties are,

1. That every strait row (horizontal or vertical) of 8 numbers added together, makes 260, and half of each row half 260.

In the modern lingo, we would call the vertical pattern a "column," not a row. Franklin's explanations will be easier to understand if we insert a few illustrations. Here are some row, column, half-row, and half-column groupings, laid out like a search-a-word puzzle:

2. That the bent row of 8 numbers, ascending and descending diagonally, *viz.* from 16 ascending to 10, and from 23 descending to 17;

52	61	4	13	20	29	36	45
14	3	62	51	46	35	30	19
53	60	5	12	21	28	37	44
11	6	59	54	43	38	27	22
55	58	7	10	23	26	39	42
9	8	57	56	41	40	25	24
50	63	2	15	18	31	34	47
16	1	64	49	48	33	32	17

52	61	4	13	20	29	36	45
14	3	62	51	46	35	30	19
53	60	5	12	21	28	37	44
11	6	59	54	43	38	27	22
55	58	7	10	23	26	39	42
9	8	57	56	41	40	25	24
50	63	2	15	18	31	34	47
16	1	64	49	48	33	32	17

and every one of its parallel bent rows of 8 numbers, make 260. — Also the bent row from 52, descending to 54, and from 43 ascending to 45;

52	61	4	13	20	29	36	45
14	3	62	51	46	35	30	19
53	60	5	12	21	28	37	44
11	6	59	54	43	38	27	22
55	58	7	10	23	26	39	42
9	8	57	56	41	40	25	24
50	63	2	15	18	31	34	47
16	1	64	49	48	33	32	17

52	61	4	13	20	29	36	45
14	3	62	51	46	35	30	19
53	60	5	12	21	28	37	44
11	6	59	54	43	38	27	22
55	58	7	10	23	26	39	42
9	8	57	56	41	40	25	24
50	63	2	15	18	31	34	47
16	1	64	49	48	33	32	17

and every one of its parallel bent rows of 8 numbers, make 260. — Also the bent row descending from 45 to 43 descending to the left, and from 23 to 17 descending to the right, and every one of its parallel bent rows of 8 numbers make 260.

52	61	4	13	20	29	36	45
14	3	62	51	46	35	30	19
53	60	5	12	21	28	37	44
11	6	59	54	43	38	27	22
55	58	7	10	23	26	39	42
9	8	57	56	41	40	25	24
50	63	2	15	18	31	34	47
16	1	64	49	48	33	32	17

52	61	4	13	20	29	36	45
14	3	62	51	46	35	30	19
53	60	5	12	21	28	37	44
11	6	59	54	43	38	27	22
55	58	7	10	23	26	39	42
9	8	57	56	41	40	25	24
50	63	2	15	18	31	34	47
16	1	64	49	48	33	32	17

—Also the bent row from 52 to 54 descending to the right, and from 10 to 16 descending to the left, and every one of its parallel bent rows of 8 numbers make 260.

52	61	4	13	20	29	36	45
14	3	62	51	46	35	30	19
53	60	5	12	21	28	37	44
11	6	59	54	43	38	27	22
55	58	7	10	23	26	39	42
9	8	57	56	41	40	25	24
50	63	2	15	18	31	34	47
16	1	64	49	48	33	32	17

52	61	4	13	20	29	36	45
14	3	62	51	46	35	30	19
53	60	5	12	21	28	37	44
11	6	59	54	43	38	27	22
55	58	7	10	23	26	39	42
9	8	57	56	41	40	25	24
50	63	2	15	18	31	34	47
16	1	64	49	48	33	32	17

—Also the parallel bent rows next to the above-mentioned, which are shortened to 3 numbers ascending, and 3 descending, &c. as from 53 to 4 ascending, and from 29 to 44 descending, make, with the 2 corner numbers, 260.—Also the 2 numbers 14, 61 ascending, and 36, 19 descending, with the lower 4 numbers situated like them, *viz.* 50, 1, descending, and 32, 47, ascending, make 260.—And, lastly, the 4 corner numbers, with the 4 middle numbers, make 260.

52	61	4	13	20	29	36	45
14	3	62	51	46	35	30	19
53	60	5	12	21	28	37	44
11	6	59	54	43	38	27	22
55	58	7	10	23	26	39	42
9	8	57	56	41	40	25	24
50	63	2	15	18	31	34	47
16	1	64	49	48	33	32	17

52	61	4	13	20	29	36	45
14	3	62	51	46	35	30	19
53	60	5	12	21	28	37	44
11	6	59	54	43	38	27	22
55	58	7	10	23	26	39	42
9	8	57	56	41	40	25	24
50	63	2	15	18	31	34	47
16	1	64	49	48	33	32	17

52	61	4	13	20	29	36	45
14	3	62	51	46	35	30	19
53	60	5	12	21	28	37	44
11	6	59	54	43	38	27	22
55	58	7	10	23	26	39	42
9	8	57	56	41	40	25	24
50	63	2	15	18	31	34	47
16	1	64	49	48	33	32	17

So this magical square seems perfect in its kind. But these are not all its properties; there are 5 other curious ones, which, at some other time, I will explain to you.

As far as we can tell, he never did so; perhaps further elaboration occurred in a letter that does not survive, or else they discussed the matter in person during their visits together in the 1760s.[15] But we might speculate on what the other curious properties are, and in fact there are far more than five of them left to consider.

For example, the design above at left can be shifted up or down, or flipped upside down, and it still totals 260. And it is not likely to be an accident that Franklin's "bent rows" can be shifted all the way off the edge of the square and continued around the other side, without losing their magical sum of 260 (figure 5.6). Moreover, the bent rows can be stretched out, like knight's-move patterns on a chessboard, and the sum is still, you guessed it, 260! (figure 5.7)

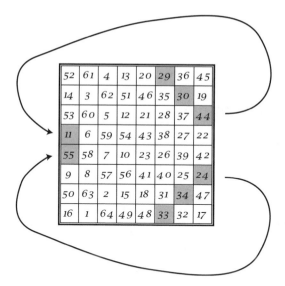

Fig. 5.6.

Fig. 5.7.

Each of these designs can be shifted into others, even falling "off the edge of the earth," to get twenty-eight more magical patterns.

There are many, many others, and you can make sport of looking for them yourself.[16] At the heart of many of these is the simple fact that, astoundingly, every 2×2 block of cells in Franklin's square sums to 130, or half the magic 260. By adding and subtracting such blocks carefully, you will unearth a myriad of patterns. It's as if Franklin balanced his square like a delicate equation, carefully ensuring that each little fold of the map weighs exactly the same. Indeed, as we have already observed, physical balance is inherent in this magic square.

I suppose we can all agree that Franklin had successfully met the challenge, and had exceeded all reasonable expectations. But in mathematics there is always a higher mountain to climb:

> Mr. *Logan* then shewed me an old arithmetical book, in quarto, wrote, I think, by one *Stifelius*, which contained a square of 16, that he said he should imagine must have been a work of great labour; but if I forget not, it had only the common properties of making the same sum, *viz.* 2056, in every row, horizontal, vertical, and diagonal.

This *Stifelius* was the German priest Michael Stifel, whose unorthodox views included a pseudo-mathematical argument that Pope Leo X was the Antichrist, based on a numerological analysis interpreting the pontiff's name as 666.[17] Applying similar methods to Biblical text, Stifel deduced that the end of the world was due on October 19, 1533. When the 20th of October rudely materialized in contravention of his prophecy, this friend and follower of Martin Luther moved on to more serious mathematics, at least for a while. His *Arithmetica Integra* (1544) was a landmark book in the history of mathematics.[18] Stifel popularized symbols like $+$, $-$, and $\sqrt{}$, without which your hand calculator would look considerably different. He presented a version of "Pascal's triangle" more than a century before Pascal himself did so (though long after its debut in Asia). His work

Fig. C.1. *Benjamin Franklin*, by David Martin (after Peale), 1767. American Philosophical Society Library.

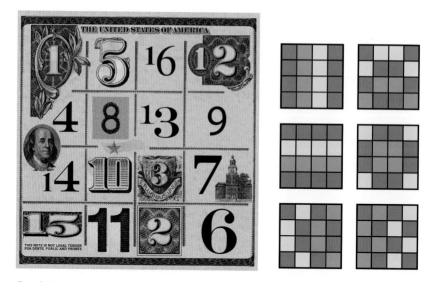

Fig. C.2. A simplified Franklin-style magic square. The first sixteen counting numbers are displayed in a carefully planned array, so that rows, columns, and many other configurations have the same "magic sum" of 34. A few of the magic sums in this particular diagram can be found by tracing the color-coded patterns at right.

Fig. C.3. Commemorative postage stamp showing magic square (detail on next page). *Benjamin Franklin, Scientist* © 2006 United States Postal Service. All Rights Reserved. Used with permission.

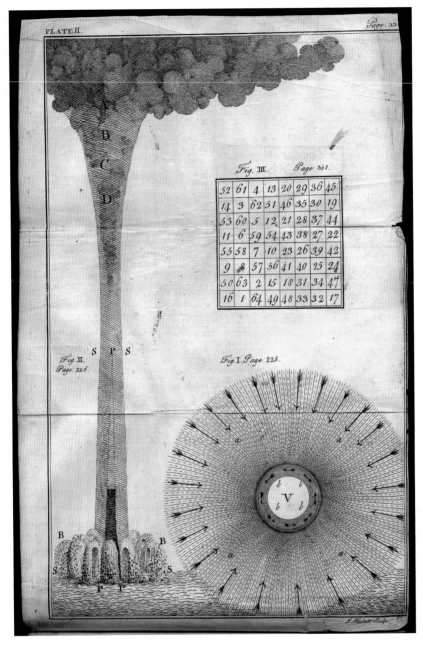

Fig. C.4. A page from the 1769 edition of Franklin's *Experiments and Observations on Electricity* showing the best-known magic square, with two views of a whirlwind. American Philosophical Society Library.

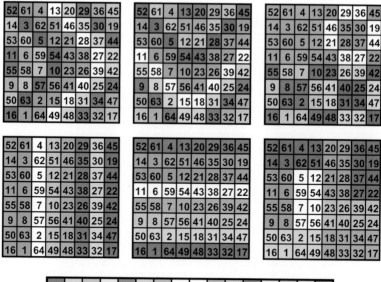

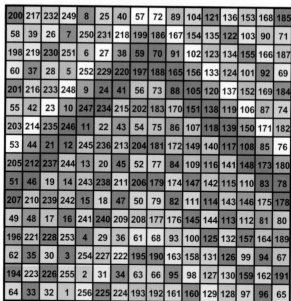

Fig. C.5. Magic patterns in Franklin's two published squares. Each color-coded pattern in the 8×8 square adds to 260. Each pattern in the 16×16 square adds to 2056.

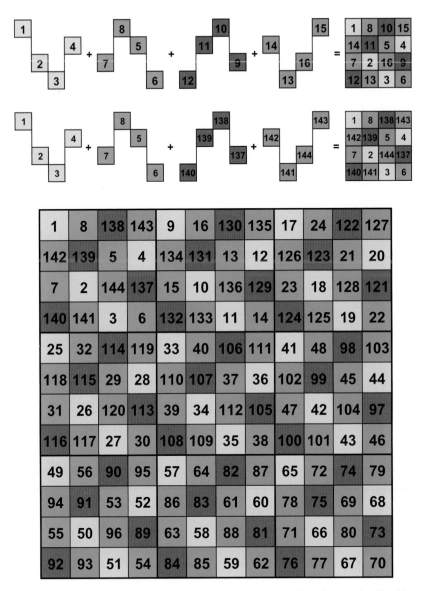

Fig. C.6. The "patchwork quilt" method from an unsigned author in the Franklin papers (chapter 6). Each row, column, and diagonal adds to 870.

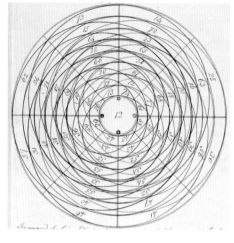

Fig. C.7. Franklin's *Magic Circle of Circles*, handwritten in color with the aid of a compass, *Canton Papers of the Royal Society,* © The Royal Society. Below: a colorized version, typeset to mimic the published black-and-white rendering sanctioned by Ferguson's *Tables and Tracts.*

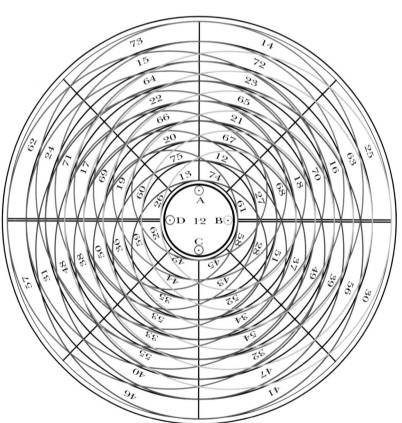

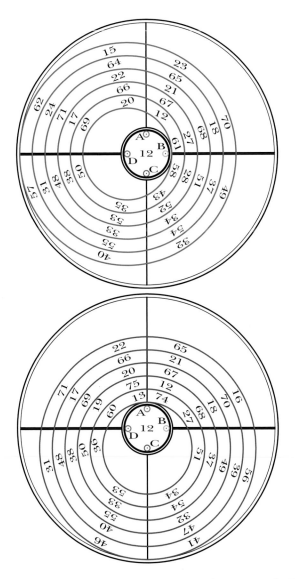

Fig. C.8. Selected "excentric" circles from Franklin's magic circle. The sum of eight entries along any circle, together with the central 12, makes 360.

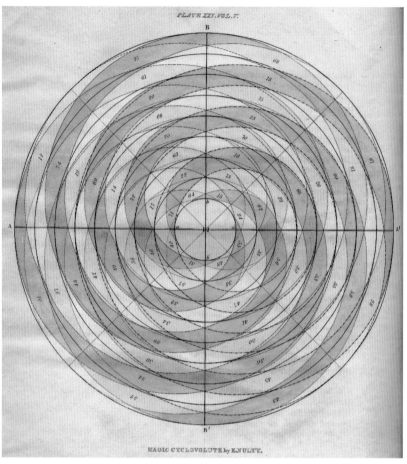

MAGIC CYCLOVOLUTE by E.NULTY.

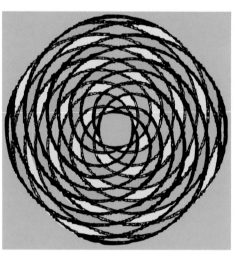

Fig. C.9. Top: Eugenius Nulty's *magic cyclovolute*, an homage to Franklin's magic circle. (*Transactions of the APS, New Series*, Vol. 5, p. 208). American Philosophical Society Library. Left: the spirals alone.

helped motivate the invention of logarithms by the later Scottish mathematician John Napier, who shared Stifel's penchant for conspiracy: this time Pope Clement VIII was the Antichrist, and the universe would conclude by 1700.[19] Aside from addressing a variety of mathematical topics, such as numerical sequences, exponentiation, and even music, *Arithmetica Integra* boasted magic squares as large as 9×9 and 16×16, though nothing numerologically catastrophic is attributed to these figures. But, as noted, Stifel's magic squares were the generic sort. With only straight diagonals to boast of, they are what Franklin considers to be the vanilla of magic squares.[20]

Not willing to be out-done by Mr *Stifelius*, even in the size of my square,[21] I went home, and made, that evening, the following magical square of 16, which, besides having all the properties of the foregoing square of 8, *i.e.* it would make the 2056 in all the same rows and [bent] diagonals, had this added, that a four square hole being cut in a piece of paper of such a size as to take in and shew through it, just 16 of the little squares, when laid on the greater square, the sum of the 16 numbers so appearing

200	217	232	240	8	25	40	
58	39	26	7	250	231	218	1
198	219	230	251	6	17	38	
60	37	28	5	252	229	220	1
201	216	233	248	9	24	41	
55	42	23	10	247	234	215	
203		235	246	11	22	43	

through the hole, wherever it was placed on the greater square, should likewise make 2056.[22]

This I sent to our friend the next morning, who, after some days, sent it back in a letter, with these words: — "I return to thee thy astonishing or most stupendous piece of the magical square, in which" — but the compliment is too extravagant, and therefore, for his sake, as well as my own, I ought not to repeat it. Nor is it necessary; for I make no question but you will readily allow this square of 16 to be the most magically magical of any magic square ever made by any magician. [See figure 5.8.][23]

It's just as well that Logan's comments are cut short, for "astonishing or most stupendous" is already compliment enough! It is often implied that this quotation comes to us directly from a Logan document, when (as you can see) it is actually a secondhand recounting from a hardly disinterested source, citing a dead man who can hardly protest.[24] As it happens, however, there is some corroboration for these sentiments elsewhere.

I have always wondered what happened to Logan's letter. Much correspondence from that era is lost, even communications between such eminent individuals. Often these writers mention that a prior letter has gone astray, disappeared in the post somewhere during their journeys by sea or land. Many other letters that were received by their addressees have vanished since then. Collinson's response is not on record, for example, and the note that returned the magic square to its inventor is also not in evidence. (That Collinson did reply there can be no doubt, for Franklin writes some time later: "I am glad the perusal of the magical squares afforded you any amusement."[25]) To compound matters, whereas some authors, like Franklin and Logan, kept drafts of their correspondence, Collinson claims not to have done so.[26] So, just what was Logan's reaction? Did he really wax so admiring over the numerical curiosities? In general, how

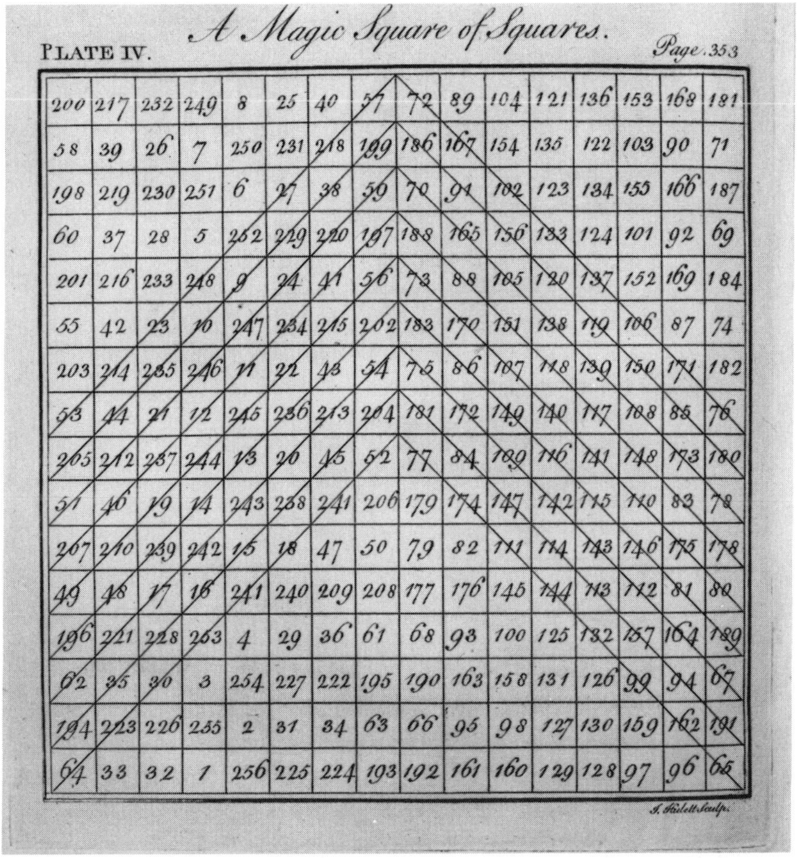

200	217	232	249	8	25	40	57	72	89	104	121	136	153	168	181
58	39	26	7	250	231	218	199	186	167	154	135	122	103	90	71
198	219	230	251	6	27	38	59	70	91	102	123	134	155	166	187
60	37	28	5	252	229	220	197	188	165	156	133	124	101	92	69
201	216	233	248	9	24	41	56	73	88	105	120	137	152	169	184
55	42	23	10	247	234	215	202	183	170	151	138	119	106	87	74
203	214	235	246	11	22	43	54	75	86	107	118	139	150	171	182
53	44	21	12	245	236	213	204	181	172	149	140	117	108	85	76
205	212	237	244	13	20	45	52	77	84	109	116	141	148	173	180
51	46	19	14	243	238	241	206	179	174	147	142	115	110	83	78
207	210	239	242	15	18	47	50	79	82	111	114	143	146	175	178
49	48	17	16	241	240	209	208	177	176	145	144	113	112	81	80
196	221	228	253	4	29	36	61	68	93	100	125	132	157	164	189
62	35	30	3	254	227	222	195	190	163	158	131	126	99	94	67
194	223	226	255	2	31	34	63	66	95	98	127	130	159	162	191
64	33	32	1	256	225	224	193	192	161	160	129	128	97	96	65

Fig. 5.8. "The most magically magical square ever made by any magician." *A Magic Square of Squares,* Plate IV, p. 353, from Franklin's *Experiments and Observations on Electricity*, 1769, The Historical Society of Pennsylvania. Nine bent rows are outlined explicitly. A careful comparison with the 8-square reveals that both examples are built according to the same recipe.

much of Franklin's account is remembered accurately? Later commentators have taken him at his word, but the famous letter from Franklin to Collinson is known mainly through a version that was published seventeen years after the fact, time enough for some creative editing. The original is gone, and the only extant draft is fragmentary.

Luckily, some of Logan's records survive, though most of their contents have never appeared in print. These do cast further light on the magical events and their timing, and make it clear that Franklin was not exaggerating the impression he made on his mentor. The image emerges that Logan was an active cheerleader for Franklin, extolling the talents of this great inventor and scientist again and again. To Collinson he relays frequent testaments to Franklin's many abilities. Much of it reads like nothing more than a long-running reference letter. In 1749 he writes of "our most ingenious Printer and Postmaster Benj Franklin who has the clearest understanding with an extream modesty of any man I know here. . . ."[27] A few weeks later Logan avers that his colleague "has a most excellent Judgem[ent] equall'd by his Modesty." He goes on to praise Franklin's potential as a mathematician in the highest possible terms, opining that with sufficient study, "he would soon equal himself almost to the greatest in that way." At risk of repeating himself, he writes again to Collinson soon after:

> Our Benj. Franklin is certainly an extraordinary Man in most Respects one of a singular good Judgment but of equal Modesty. He is Clark of our Assembly and there for want of other Employment while he sate Idle he took it in his Head to think of Magical Squares in which he out did Frenicle himself who published above 80 pages in Folio on that Subject alone which I have amongst Divers Ouvrages de Mathematique &c. . . .

On the final pages of this handwritten volume—Logan was in ill health now and would die the following year—are two letters concerned with our protagonist. The first is to Collinson again:

> Feb 28 1750.[28] Dear Peter, I have spent most of this day for the first time with thy friend Kalm [a Swedish botanist] accompanied with B. Franklin. . . . Pray do not imagine I overdoe it in my Character of BF for I am rather short in it and I hope to convince thee when an opportunity offers free of Postage that what I wrote in my last by Capt James of his Magical Squares is truly Astonishing.

And, we might add, most stupendous. The other is addressed to his bookseller and frequent correspondent, Jonathan Whiston. (This was the son of Newton's successor at Cambridge, William Whiston, one of Poor Richard's sources.[29]) The entry begins, "[Mar 1 1750.] Being yesterday in company with B. F. our most ingenious Printer and now our Postmaster...." If Logan's English connections remained unconvinced of Franklin's genius, it would not be through lack of effort on the part of this outspoken advocate.

These comments and others in the same letter-book finally provide a rough timeline for the events described in this chapter. An entry for June of 1749 says simply, "Wrote to B. Franklin to come up and See my Books." We know from other records that Franklin visited Stenton in late August or early September, and possibly on many other occasions around the same time. The most telling item:

> [Jan 12 1750] I wrote to him to come up hither next first day if the weather was good Seeing while the Assembly Sits I can app[oint] no other day and if my Son has delvd him the Magic Squares I pray him to bring them with him.

Thus, these magical events transpired while Franklin was in the midst of his electrical experiments. Apparently those world-shattering physical investigations did not occupy his mind to the exclusion of all other intellectual matters.

Having matched and even surpassed his French and German predecessors with two glorious magic squares, and thus having satisfied friend Logan's challenge, Franklin could have called off the hunt. Such unsporting behavior would not befit an aspiring mathematician, however. He says as much to Collinson:

> I did not, however, end with squares, but composed also a magick circle, consisting of 8 concentric circles, and 8 radial rows, filled with a series of numbers. . . . If you desire it, I will send it; but at present, I believe, you have enough on this subject.
>
> I am, &c. B.F.

Notes

1. The definitive source on Logan's library is Edwin Wolf 2nd, *The Library of James Logan of Philadelphia 1674–1751*, Philadelphia: The Library Company of Philadelphia, 1974. See also Wolf, "The Romance of James Logan's Books," *The William and Mary Quarterly*, 3rd series, Vol. 13, No. 3, 1956, pp. 342–353. The earliest claim regarding Newton's work is by Frederick E. Brasch, "James Logan, a Colonial Mathematical Scholar, and the First Copy of Newton's Principia to Arrive in the Colony" (see note 30 of chapter 4). Although the title intimates that he refers to Pennsylvania alone, the article itself makes clear that a broader claim is at stake (see caption, p. 6). It hardly matters; either way this can never be verified to any reasonable degree of certainty.

2. Various scholars have observed that, while at least two colonial libraries were larger in sheer size, Logan's was the finest in quality. Some of the comparisons of Franklin and Logan are noted by Frederick B. Tolles in the introduction to "Philadelphia's First Scientist: James Logan" (note 33, chapter 4).

3. Another celebrated correspondent was William Jones, who served on the committee to arbitrate the controversy over whether Newton or Leibniz should be credited with discovering the calculus. He would also become embroiled in the Godfrey-Hadley priority dispute. Roy N. Lokken, "The Scientific Papers of James Logan," *Transactions of the APS* (new series), Vol. 62, No. 6, 1972, pp. 1–94; also, Tolles, cited above.

4. Roy N. Lokken, "The Social Thought of James Logan," *The William and Mary Quarterly*, 3rd series, Vol. 27, No. 1, 1970, pp. 68–89.

5. These quotes are from Tolles, cited above; Brasch, cited above; and Edwin Wolf 2nd, in Charles Coulston Gillispie, ed., *Dictionary of Scientific Biography*, Vol. VIII, American Council of Learned Societies, New York: Charles Scribner's Sons, 1973.

6. Likewise Logan was the single most important influence on Godfrey, who (like Flamsteed, Newton, Huygens, and Halley) was occupied with the longitude problem. According to legend, Logan entered his famous library one day to find an uninvited visitor perusing a volume of Newton. This was Godfrey, who would learn mathematics on his own and go on to make his great discovery. The better attested version of the story has him knocking at the door to ask permission first. Either way, Logan, through his books and encouragement, is indirectly responsible for Godfrey's invention.

7. *Experiments and observations*, 1769 (see note 60 of chapter 4 for complete citation), pp. 1–2.

8. Historian of science I. Bernard Cohen, author of *Benjamin Franklin's Science* and *Franklin and Newton* (and editor of the twentieth-century edition of the *Experiments*) averred that, contrary to today's image, Franklin's scientific reputation was by far the larger part of his fame. It provided him with credibility and "political capital." In the review mentioned earlier (our chapter 1, note 10), for example, Cohen notes that "much of Franklin's political success on the continent was due to the reputation he had established as a scientist."

9. When I went to Stenton for the first time, I was the only visitor that day. It is a historic site that deserves to be better known. At different times, both the American and British commanders used the house as a headquarters, prior to the Battle of Germantown.

10. It was "in the country" then, but the city of Philadelphia has grown around Stenton since.

11. This and later excerpts are taken from Benjamin Franklin, *Experiments and observations . . .* , 1769.

12. Academie Royale des Sciences: *Divers Ouvrages de Mathematiques et de Physique*, Paris: De l'Imprimerie Royale, 1693, p. 429.

13. Many historical sources have misattributed the allusion to Frénicle, claiming Franklin refers here to *Traité des triangles rectangles en nombres, dans lequel plusieurs belles proprietés de ces triangles sont demontrées par de nouveaux principes*, 1676. (*Papers* 1961, Vol. 4; *Writings*, Vol. 2; Cohen, *Benjamin Franklin's Science*; and Frank J. Swetz, *Legacy of the Luoshu: The Mystical, Mathematical Meaning of the Magic Square of Order Three*, Chicago: The Open Court Publishing Company, 2002.) But while that work is number-theoretic, it contains no magic squares. In any case it was not part of Logan's library; see both Wolf references in note 1. The correct reference is given there; his bibliographer confirms that *Divers ouvrages* was in Logan's possession.

14. It is difficult to imagine that the owner of such an extensive collection of Tipper's almanacs really believed this to be the case. Indeed, Franklin's own short-lived *General Magazine* was also guilty of such indulgence.

15. That there were many such visits is well attested by their surviving correspondence. Also, "Visit to Peter Collinson in 1767 by Benjamin Franklin," Carlotte H. Browne, MSS Case 3, The Historical Society of Pennsylvania.

16. Pasles, 2001, and Pasles, "Franklin's Other 8-Square," *Journal of Recreational Mathematics,* Vol. 31, No. 3, 2003.

17. That *Stifel* Latinized to *Stifelius* is nothing notable, and the latter name adorns the spine of Logan's volume as Franklin recalls. Meanwhile the title page has him as *Stifelio*, and the running header throughout shows *Stifelii*.

18. Michaele Stifelio, *Arithmetica Integra*, Nürnberg: Iohan. Petreium, 1544.

19. Our amusement should be tempered by the seriousness with which apocalyptic visions are often held in our own enlightened age.

20. Michael Stifel's magic squares can be seen at http://www.pasles.org/magic.html.

21. He misremembers slightly, or else they did not look very closely at Frénicle, for the latter's work includes a 14-square that can be extended to still larger orders. Thus the 8-square had already been "outdone" in size before the two men leafed through *Arithmetica Integra*.

22. More specifically, every 2×2 block of cells totals 514. Franklin's description of this property is strongly reminiscent of "grille" cryptography, in which a card is placed over another sheet so that only certain letters or numbers remain visible. (W. W. Rouse Ball and H.S.M. Coxeter, *Mathematical Recreations and Essays*, 13th edition, New York: Dover, 1987.)

23. His diagram highlights bent rows that fall within the boundary of the square but not those that cross the edge, which is why I have categorized these separately before. It appears that Franklin does not explicitly count the edge-crossers among his "parallel bent rows," yet he must have been aware of them.

24. See, for example, *James Logan, 1674–1751: Bookman Extraordinary,* Philadelphia: Library Company of Philadelphia, 1971.

25. For that matter, we also have no record of a letter that Franklin may reference in his preamble ("According to your request, I now send you the Arithmetical Curiosity . . ."), though perhaps that request was forwarded by Logan or some other third party.

26. Letters of Collinson to Franklin, Aug. 15 & Sept. 27, 1752, *Papers*, 1961, Vol. 4, pp. 341–342 and 357. Collinson mentions some lost correspondence, which may include his response to the magic squares.

27. This and subsequent quotations of James Logan are from his Letter-book, 1748–1750, Logan Papers No. 5, The Historical Society of Pennsylvania.

28. As elsewhere in this book, dates have been standardized here according to the modern calendar. For example Logan actually writes the year as $17\frac{49}{50}$, since the English had not yet made a complete transition to the Gregorian calendar; and months are counted according to the Old Style, whence 1 = March, 2 = April, and so on. (Incidentally, this numbering explains the etymology of Sept., Oct., Nov., and Dec.)

29. Whiston the elder held the Lucasian chair until he was fired for heresy. On the cover of Franklin's first almanacs, one will find the year calculated according to "the computation of W. W." He is cited more explicitly in *Poor Richard* for 1747 and 1748.

6

The Mutation Spreads
(Adventures Among the English)

> The magic square and circle, I am told, have occasioned
> a good deal of puzzling among the mathematicians here;
> but no one has desired me to show him my method of
> disposing the numbers. It seems they wish rather to
> investigate it themselves.
> —Benjamin Franklin, 1768

In 1767, James Ferguson's *Tables and Tracts, Relative
to Several Arts and Sciences* was printed in London, bringing with
it the first published appearance of Franklin's magic 16-square.[1]
Current accounts would have you believe that no one followed
up on the great inventor's magical creations until very recently.
But my own excursion through scores of tremendously rare books
and manuscripts has unearthed a number of authors who fur-
thered Franklin's magic squares and circle in the decades following their initial publication. (The magic circle will be revealed in
chapter 7.)

Like its biological counterparts, a meme evolves over time.[2] The
magic square meme was particularly rejuvenated by Franklin's art-
ful genetic manipulations, and that effect was felt first in the City of
Brotherly Love.

In the year after their public unveiling, Franklin's magic square
and circle began to arouse interest among his fellow colonists,

specifically as a topic of discussion at a meeting of the American Philosophical Society on June 21, 1768.[3] Members of the Society gathered at the State House, the future Independence Hall, where Franklin the Assembly clerk had once idly doodled his squares and circles. Later, the First and Second Continental Congresses would convene here. Indeed, this was precisely eight years to the day before Thomas Jefferson asked Franklin to improve his draft, and Franklin would write to George Washington "that a Declaration of Independence is preparing."[4]

At that meeting, a mathematical paper was read by one of the Society's secretaries (figure 6.1). This was the Reverend John Ewing, a man of no mean resumé himself. Born in Maryland near the Pennsylvania border, he would one day serve on the commission to precisely define that boundary, a part of the Mason-Dixon Line that would divide the imperiled Union first joined in this hallowed Hall. His colleague and biographer, Robert Patterson—whose navigation lessons guided the Lewis and Clark expedition—claims Ewing's mathematical prowess so exceeded his instructor's that he could be considered essentially "self-taught."[5]

Having thus outpaced his teacher, Ewing attended Princeton (then known as the College of New Jersey). There he was a favorite student of Aaron Burr, father of the future vice president. Ewing entered the College as a senior in 1754 and graduated within one year.[6]

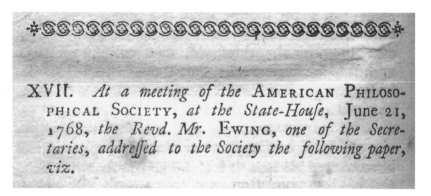

Fig. 6.1. American Philosophical Society Library.

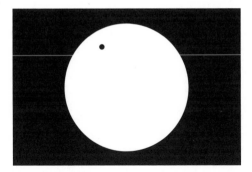

Fig. 6.2. Transit of Venus.

Together with Franklin, David Rittenhouse, and others, Ewing participated in a coordinated international effort to observe the historic transit of Venus across the face of the Sun in 1769 (figure 6.2).[7] Likewise, he recorded the transit of Mercury, a more common event, a few months later. These "tiny eclipses," too rare to bear notice in the almanacs, were of great importance to astronomers; careful measurements would make it possible to determine the distance to the Sun, enabling more accurate navigation.[8]

Mercury and Venus aside, the Reverend Ewing would hardly have entertained any ancient superstitions of the seven "planets," as assumed in Agrippa's alchemy or Poor Richard's astrology. Yet, he did find occasion to fiddle with some magic, as we shall see.

Ewing's published lectures included this prescient bit of wisdom:

> [At great interstellar distances,] our whole planetary system must become absolutely invisible to eyes such as ours. Hence it is extremely probable, that every one of this countless multitude of stars was ordained for similar purposes with our sun; and that they are the central suns of innumerable systems, distributing the various influences of light and heat to their attending planets; whose immense distance from us, must for ever conceal them from the human eye.[9]

Perhaps such an opinion was natural for someone who engaged professionally in matters both spiritual and scientific. As we have seen, Ewing possessed a varied *curriculum vitae* that would impress more, if only we were not discussing him in the shadow of Franklin's. He was an astronomer, mathematician, educator, administrator, and clergyman. And in 1768 his mind was on magic. He writes:

> Finding in a late publication of Mr. Ferguson, a Magic Square, communicated to him by Dr. Franklin, the author of the invention, the principles of which, Mr. Ferguson confesses, he did not understand; I was desirous of discovering a general method for constructing Magic Squares of any given dimension, which the ingenious inventor thought proper to conceal. Having succeeded in the attempt, I beg leave to lay a Magic Square before the Society, constructed on the same principles with that of Dr. Franklin.[10]

Unfortunately, the *Transactions* for that day records only a convoluted, indirect set of instructions for building Ewing's square, and no concrete example is actually displayed for our inspection. He left a vague description so riddled with errors and (possibly deliberate) obfuscations that when the Society considered publishing Ewing's square, seventy-six years after the fact, the consulting mathematician found it impossible to correctly interpret his directions and the proposal was abandoned indefinitely.[11] In essence, Ewing accuses Franklin of publishing an example with no method, and responds by publishing a method with no real example! However, I have studied this paper with the benefit of modern technology, and have successfully divined his intent. Thus I beg leave to display for you, finally, the Magic Square of John Ewing (figure 6.3, right).

This is the conventional sort of magic square: rows, columns, and diagonals sum to 3420. But it's also more than that. Strip away the outermost layer, so that it now begins 60, 61, 62, and so on. The result is again a magic square! (The rows, columns, and diagonals of this smaller diagram all total to 3060.) Such an animal is known today as a *border square*. Borders are a recurring theme in Ewing's life story. In fact, his creation is "multiply bordered": as you peel away one layer after another, you get a new magic square each time.[12]

Ewing's writings indicate that his method can be used to build a magic square of order 3, or 5, or 7, and so on, but he chooses to focus on a 19×19 grid for the occasion. Why? There is a popular but misguided belief that a larger square automatically represents a more difficult achievement, akin to the schoolchild's prediction that higher-level math must concern bigger numbers. A square of size 17 would make it too obvious that he is competing with Franklin's 16, and 19 is the next odd number after 17. (Ewing's particular technique would not work for an even size, like order 18, without further adaptations. [13]

It's a marvelous creature, this magic square, but it lacks the myriad qualities associated with Franklin's masterpieces. Thus, I fear that his best effort did not match Franklin's own, though there is no shame in falling short of the gold standard.

The aftermath (pardon the pun) was somewhat contentious.

An account of Ewing's remarks appeared on the front page of Franklin's old newspaper, now run by his successor and ex-partner David Hall. The tortuous description of this magic square began just below the banner *Pennsylvania Gazette: Containing the Freshest Advices, Foreign and Domestic*. The rest of the page was taken up by announcements: reward for a prisoner escaped from the Trenton jail ("speaks French and high Dutch, but indifferent English"); announcements of goods for sale (a carriage, a slave).[14]

The immediate response was an angry one; some people take mathematics *very* seriously. The *Gazette*'s competitor, the *Pennsylvania Chronicle*, ran a letter indignantly accusing the reverend mathematician of casting doubt on Dr. Franklin's forthrightness. Mr. Ewing "insidiously hints the ingenious Inventor had thought proper to conceal" his general method, says the pseudonymous correspondent:[15]

Now Sir, as it is well known to all the Doctor's Acquaintance with what Freedom he communicates any Discoveries that he makes, by giving th[is] following Extract from Mr. *Ferguson's* Tracts a Place in your next Chronicle, the Public will then be enabled to judge of the Candour and Ingenuity of the REVEREND SECRETARY. . . .

Fig. 6.3. Above: *John Ewing* by Charles Willson Peale,
1788, National Portrait Gallery, Smithsonian Institution;
partial gift of Dean Emerson and Maisie Emerson Macy.
A telescope is shown within easy reach. Peale,
renowned portrait artist and naturalist, also painted
Franklin, Washington, Rittenhouse, and Benjamin Rush.
Facing page: The numbers from 0 to 360 arranged so
that, no matter how many outer layers are stripped
away, the result is always a magic square.

"What the Doctor's Rules are for disposing of the different numbers
so as that they shall have the following Properties, I know nothing
of, and perhaps the Reason may be, that I have not ventured to ask
him; ALTHOUGH I NEVER SAW A MORE COMMUNICATIVE MAN IN
MY LIFE."

27	28	29	30	31	32	33	34	35	18	343	344	345	346	347	348	349	350	351
360	60	61	62	63	64	65	66	67	52	309	310	311	312	313	314	315	316	0
359	324	89	90	91	92	93	94	95	82	279	280	281	282	283	284	285	36	1
358	323	292	114	115	116	117	118	119	108	253	254	255	256	257	258	68	37	2
357	322	291	264	135	136	137	138	139	130	231	232	233	234	235	96	69	38	3
356	321	290	263	240	152	153	154	155	148	213	214	215	216	120	97	70	39	4
355	320	289	262	239	220	165	166	167	162	199	200	201	140	121	98	71	40	5
354	319	288	261	238	219	204	174	175	172	189	190	156	141	122	99	72	41	6
353	318	287	260	237	218	203	192	179	178	183	168	157	142	123	100	73	42	7
352	317	286	259	236	217	202	191	184	180	176	169	158	143	124	101	74	43	8
26	59	88	113	134	151	164	173	177	182	181	187	196	209	226	247	272	301	334
25	58	87	112	133	150	163	170	185	188	171	186	197	210	227	248	273	302	335
24	57	86	111	132	149	159	194	193	198	161	160	195	211	228	249	274	303	336
23	56	85	110	131	144	207	206	205	212	147	146	145	208	229	250	275	304	337
22	55	84	109	125	224	223	222	221	230	129	128	127	126	225	251	276	305	338
21	54	83	102	245	244	243	242	241	252	107	106	105	104	103	246	277	306	339
20	53	75	270	269	268	267	266	265	278	81	80	79	78	77	76	271	307	340
19	44	299	298	297	296	295	294	293	308	51	50	49	48	47	46	45	300	341
9	332	331	330	329	328	327	326	325	342	17	16	15	14	13	12	11	10	333

Fig. 6.3. (continued)

Such is the ingenious Mr. *Ferguson's* Opinion of our great AMERICAN PHILOSOPHER, whose Name and Virtues will be handed down with Honour to Posterity, when the Reverend Secretary, and a thousand other little carping Momus's shall have sunk into their *original Nothingness*.

CANDIDUS.

(In case your own classical education is only a distant memory, Momus is the Greek god of unfair criticism.[16]) And that seems to be the end of the exchange, at least in print; when Ewing's ideas were recapitulated in the *American Magazine* the following year, there was no outcry or rejoinder this time.[17] Whether Franklin himself felt insulted is not known. However, in November of that year, he received a letter implying that Ewing and his colleagues had taken credit due to others for observing properly the transit of Venus:

> A Pirate, and a Highwayman, I feel some charity for; but a lyar and filcher of reputation, are to me, the most detestable scunks in human society.[18]

But Franklin resisted the urge to respond with similar vituperation. His reply, devoid of animosity, displays instead his characteristic diplomacy. He politely thanks the correspondent for his letter without taking a position, but mentions that he has also been corresponding with Ewing. There is certainly no evidence of a personal affront, whatever his unnamed proxy might have opined.

Another mathematician's experiments with magic squares are recorded in Franklin's papers, on two sides of a single sheet that probably dates to the 1760s or 1770s.[19] Franklin's greatest biographer (after himself), Carl Van Doren, alleges that this unsigned page "gives Franklin's rule for making magic squares,"[20] but what it describes does not correspond to the examples we have seen, and it was probably invented by someone else. (Like many of the papers in that collection, its author is not identifiable.[21]) Because it is only outlined briefly, this method has never been published before, but it can really be explained very easily.[22] To avoid even the simple algebraic notation used by "Anonymous," I will use an entirely different, color-coded system instead to make it easier to follow.

1	8	10	15
14	11	5	4
7	2	16	9
12	13	3	6

The easiest way to understand the method of Anonymous is to begin with the perfectly innocent 4-square shown above, item 94 in Frénicle's catalog. We might deconstruct this magic square as follows. Think of the whole numbers from 1 to 16 as being grouped into quartets: 1,2,3,4, then 5,6,7,8, then 9,10,11,12, and finally

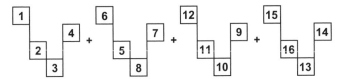

Fig. 6.4.

13,14,15,16. Arrange each quartet as shown in figure 6.4.[23] By rotating and flipping these configurations in the proper way, they can be fitted together into a magic square, overlaid like interlocking puzzle pieces. See figure 6.5, top. (A four-color version of the process can be seen in the color plates.)

Now, if you wanted to create a bigger square, of size 8, 12, 16, or any other multiple of 4, Anonymous claims that you could build it from 4×4 blocks patterned in the same style. Let us say your goal is a 12×12 magic square. The ingredients are the whole numbers from 1 to 144. The first couple of quartets are still 1,2,3,4 and 5,6,7,8. This time, however, the last two quartets are 137,138,139,140 and 141,142,143,144. Fit those pieces together as before (figure 6.5, middle).

Now we're left with the numbers from 9 to 136. Use the first two quartets (9,10,11,12 and 13,14,15,16) and the last two (129,130, 131,132 and 133,134,135,136) in the same fashion to make another 4×4 magic square. Do this again and again, using up sixteen numbers each time, until you have exhausted all of the numbers from 1 to 144 and have nine magic squares to show for your troubles. These nine are the bricks used to build the final result, a patchwork quilt of nine tiles that can be sewn together in any order you like, for example as in figure 6.5, bottom. (*As an exercise, use the Anonymous method to construct an 8×8 magic square. Do its magic properties depend on the order in which the 4×4 tiles are quilted together?*[24])

Between them, Ewing and Anonymous covered many possible sizes: all of the odd numbers, and all multiples of 4. That leaves a few sizes unattended yet. But don't worry—Ben Franklin himself tackles those remaining cases in our chapter 8.

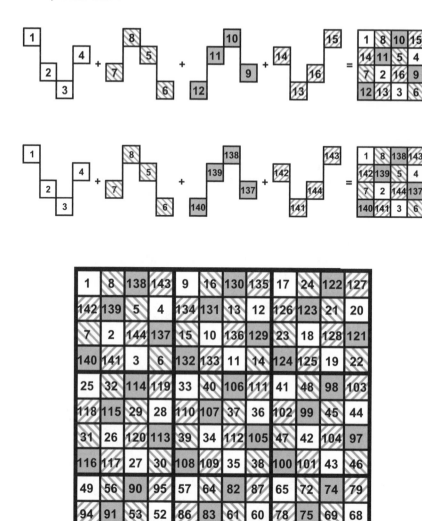

Fig. 6.5. Anonymous square.

So, who was Anonymous really? He refers to *magical* (not *magic*) squares, as Franklin often did. On the other hand, the second page of this never-published note shows a failed attempt to construct a border square, which points to Ewing instead. All in all, this is a curious document. It is unsigned, and the handwriting resembles neither suspect's. Yet it has a peculiar, labored quality about it, as if the writer was uncomfortable holding a pen, or was unused to printing carefully, or was intent on disguising his identity. For example, the letter "e" appears sometimes as "e" and other times as "ε," not a typical habit—more often, such idiosyncrasies are a sign of forgery. While my armchair handwriting analysis proves nothing definitive, at least we can evaluate the quality of this 12×12 example: noteworthy, but not to be compared with Franklin's. As with Ewing's square, we can only count on the straight diagonals that were so derided by Franklin. For orders 8 and 16, however, this method does yield bent rows and most other properties explored in the previous chapter.

The Anonymous page suggests that Franklin was communicating with like-minded individuals outside of official channels such as society transactions or formal letters. Moreover, the method as originally composed is hardly transparent to the layperson (hence Van Doren's confusion); and if such terse instructions were prepared for Franklin's use, this obviates the argument that he was not mathematically minded and possessed only a savantlike capacity for magic square making.

Certainly there must have been other experimenters besides Ewing and his unidentified colleague who did not record their work, or whose records are now lost. But I don't mean to give the impression that colonial America was a thriving center of mathematical research into magic squares. In fact, colonial America was not a thriving center of mathematical research into anything, just yet. As one historian wrote, until the turn of the eighteenth century "the only mathematics of any consequence that had been done in the United States was Franklin's work with magic squares and a few notes by David Rittenhouse."[25] America would not become a true power in mathematics for another century.[26] Until it did, stateside mathematicians would have to content themselves with the few sporadic journals that sprang up in the interim, and magic squares did appear within their pages.[27]

Across the ocean, far from the colonial backwater, it was a different story. Here was one of the great mathematical powers. In Franklin's time, England was still flush with pride over Isaac Newton's invention of calculus, so much so that a priority dispute with the Continent continued to arouse strong nationalist feelings.

If, as Franklin had alleged, the English once showed "good sense" in avoiding questions on matters as inapplicable as the magic square, that wisdom was revised in light of Franklin's efforts. By 1768, a mere one year after the publication of Ferguson's *Tables and Tracts*, some had been tempted to experiment with Franklin's own formulation of the problem.

On July 2 of that year, Franklin wrote to John Winthrop, F.R.S., a Harvard professor and great-great-grandson of his namesake, the first governor of Massachusetts Bay Colony.[28] Winthrop would become a member of the APS the following year. While their prior correspondence had concerned subjects both Biblical and scientific, the missive in question discussed mathematical instruments to be used in astronomical measurements of the transit. Franklin, writing from London, concludes with the latest news: "The magic square and circle, I am told, have occasioned a good deal of puzzling among the mathematicians here; but no one has desired me to show him my method of disposing the numbers. It seems they wish rather to investigate it themselves."[29] How like a mathematician: it is the journey, and not the destination, that is of greatest import.

But was this mere braggadocio on Franklin's part? Or were his magic square and circle only simple puzzles worthy of his amateur status? It seems that England did indeed take an interest in decoding the magic squares. Some evidence of this research activity only surfaced in print later on.

In 1795, an avowed successor to Stone's *Mathematical Dictionary* was printed in London.[30] Its author was Charles Hutton, F.R.S., who ascended from a family of coal miners to become professor of mathematics at the Royal Military Academy. Hutton gave Franklin a prominent place in his entry on the history of magic squares, averring that "the ingenious Dr. Franklin, it seems, carried this curious speculation farther than any of his predecessors in the same way." In a much later edition Hutton even performed his own analysis of Franklin's square of 16, and having carried such an interest over so great an interval, he is likely one of the English mathematicians referred to in Franklin's letter. That they were acquainted seems possible, as both were Fellows of the Royal Society and Hutton would later acquire Franklin's collection of mathematical instruments.[31] In fact, Hutton had already inherited the editorship of the *Ladies Diary*, and he also updated Ozanam's *Recreations*; earlier versions of these sources were cited as influences on Franklin in chapter 2. It is poetic that Hutton inserted Franklin's magic square of squares in the *Recreations,* the book that may have started it all.

The aforementioned dictionary was not limited to what we would consider "mathematics" today: there are instructions on how to build a proper fortification, and a quick perusal also reveals entries like "ballistics" and "fuse," probably not part of your own freshman math curriculum. But then "A Mathematical and Philosophical Dictionary" could encompass just about anything scientific or technological in those days, when *natural philosophy* connoted a breadth of subject matter difficult to fathom in our modern era of compartmentalized knowledge and expert specialization.

Though he made no single great discovery of his own, Hutton was one of the most influential mathematicians of his time, for he compiled the standard reference dictionary then in use and crafted a comprehensive mathematics course that was implemented on both sides of the Atlantic.[32]

Isaac Dalby—like Hutton, a professor at the Royal Military Academy—was a prominent figure in the longitude calculations of the day. Dalby worked to "improve" the magic square and circle, and he came up with a complex example of his own.[33] One wonders, was this mathematician/geographer inspired, perhaps, by the resemblance between the polar longitude map and the magic circle of circles (figure 6.6)?[34]

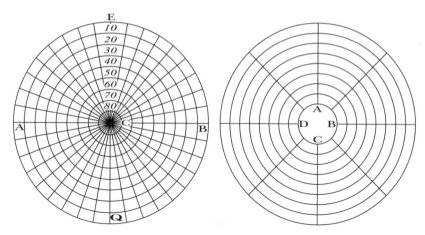

Fig. 6.6. Left: polar view of the Earth showing latitudes and longitudes: nine concentric circles cut by eighteen diameters. Right: Framework of the famous magic circle: nine concentric circles cut by eight radii. The similarity is hard to miss in Hutton's massive 1815 edition of the *Dictionary*, where the alphabetic proximity of *magic* and *maps* places these copperplates only two pages apart. Observe the radial sum of 360.

These were practical men. Hutton's dictionary covered chronometers, navigation, and military affairs. Dalby worked on practical mathematics for the Crown. Ewing was an astronomer. Yet all shared a fascination with magic squares.

Other English mathematicians who pondered and published on Franklin's work included Joseph Youle, a Sheffield schoolmaster; and the Reverend Thomas Watson, like Ewing a Presbyterian minister, but one whose main interests were decidedly nonscientific.[35] Watson's other writings bear such tantalizing titles as: "A plain statement of some of the most important principles of religion, as a preservative against Infidelity, Enthusiasm, and Immorality," while another promises "to correct the immoral tendency of some doctrines at present popular and fashionable." Like the practical mathematicians, he had weighty issues on his mind. As a mathematical amateur, he represents a species that would frequently appear among the later followers of Franklin.

What these four Britons shared was an enchantment with Franklin's magic figures and their myriad arithmetic qualities. What they also shared beyond their admiration was a respectful but unfair aesthetic objection that Franklin's figures lacked the traditional straight diagonals. Hutton, for one, refers to this as "a radical imperfection." Little did they know that the inventor himself had already answered such criticism. In a work now long forgotten, Franklin silences these carping Momuses well enough, with an improved magic square we will meet soon.

Notes

1. James Ferguson, ed., *Tables and Tracts, Relative to Several Arts and Sciences*, London: Printed for A. Millar and T. Cadell, 1767.

2. Recall from chapter 2 that a *meme* is a unit of cultural transmission.

3. John Ewing, "On Magic Squares," *The American Magazine, or, General Repository* (Jan.–Sept. 1769), appendix, publishing transactions of the American Philosophical Society XVII, 90–94, Philadelphia: William and Thomas Bradford, 1769. It should be noted that there were two distinct but like-named societies with similar purpose in Philadelphia: the American Philosophical Society (descended from Franklin's Junto), and the American Society for Promoting Useful Knowledge. In 1769 these organizations were joined.

4. Smyth, *Writings*, Vol. 6, pp. 449–450.

5. Robert Patterson, ed., in John Ewing, *A Plain Elementary and Practical System of Natural Experimental Philosophy; Including Astronomy and Chronology. With a Biographical Sketch of the Author,* Philadelphia: Hopkins and Earle, 1809. Also John Blair Linn, *A Discourse Occasioned by the Death of the Reverend John Ewing, D.D., Late Senior Pastor of the First Presbyterian Congregation, of the City of Philadelphia, and Provost of the University of Pennsylvania,* Philadelphia, 1802; and Lucy Lee Ewing, *Dr. John Ewing and Some of His Noted Connections*, Philadelphia: Press of Allen, Lane & Scott, 1924.

6. Burr the elder was also president of the college, and a Presbyterian clergyman as well.

7. While Ewing and Rittenhouse watched the transit from Pennsylvania, Captain James Cook observed from Tahiti. (Which assignment would you have chosen?)

8. The Venus event is especially rare: if you miss the next one, in 2012, you will have to wait until 2117(!) to see it again.

9. *A Plain Elementary and Practical System*, pp. 537–538. For his part, Poor Richard (1749) wrote: "It is the opinion of all the modern philosophers and

mathematicians, that the planets are habitable worlds. If so, what sort of constitutions must those people have who live in the planet Mercury?"

10. Ewing, "On Magic Squares."

11. Letter, Eugenius Nulty to Robert M. Patterson [son of Robert Patterson, the advisor to Meriwether Lewis mentioned above], Dec. 11, 1844, APS Archives, American Philosophical Society Library.

12. Border (also "bordered" or "concentric") squares were known long before Ewing's time, and in fact some examples appear in the writings of Stifel, Pascal, Frénicle, and earlier Islamic and Chinese mathematicians, going all the way back to Yang Hui's 5-square. "Nested" is another name for multiply bordered.

13. The border square method can be made to work for even squares, too, though Ewing does not discuss this possibility. W. W. Rouse Ball and H.S.M. Coxeter, *Mathematical Recreations and Essays*, 13th edition, New York: Dover, 1987.

14. *Pennsylvania Gazette*, Sept. 1, 1768.

15. *Pennsylvania Chronicle*, Sept. 5, 1768.

16. Alternately, he is the god of laughter and mockery. Momus appears in Aesop as a character who always finds the cloud behind the silver lining. His ability to see the empty half of the glass was finally frustrated: "Momus is said to have burst with chagrin at being unable to find any but the most trifling defects in Aphrodite" (1911 *Britannica*). This was still a familiar enough reference to appear in O. Henry's *The Gentle Grifter* and Carl Sandburg's *Chicago Poems* a century and a half after his appearance in the *Chronicle*. Also two of the books cited as influences in Franklin's *Autobiography* deserve mention. The preface to *Cocker's Arithmetick* (see our chapter 2) closes with this rhyme: "Zoilus and Momus, lye you down and dye, For these inventions your whole force defy." (Zoilus was a particularly harsh critic of Homer.) And Sturmy's *The Mariners Magazine* (2nd ed.) cautions "ignorant *Momus* and his Mates, who make it their business to scoff, deride, affront, and abuse all such as are Ingenious. . . . For such Loiterers there is a pair of Stocks fitted in Hell by the Devil. . . . I say, such *Momusses* will have their Heads in such Stocks. . . ."

17. *The American Magazine*, Jan.–Sept. 1769; see note 3 above.

18. Letter from Cadwalader Evans to Benjamin Franklin, Nov. 27, 1769, *Papers*, 1972, Vol. 16, p. 240.

19. Franklin Papers in the Library of the APS, XLIX, 79. American Philosophical Society Library. Like Ewing's border square, this method received brief mention in my 2001 article on Franklin; it has not been studied properly before. The date given here is based on a rough assessment of the watermark, for which I thank Robert S. Cox of the APS Library. (Personal communication, June 20, 2000.)

20. Carl Clinton Van Doren, *Benjamin Franklin*, 1938; New York: Penguin USA reprint (1991), Ch. 5, note 35. Van Doren cites Franklin Papers XLIV, 79 but that is a typographical error; he means XLIX, 79.

21. Again, I defer here to the opinion of Mr. Cox.

22. Interestingly, a similar method is described in a much later Russian source: Szczepan Jelenski, *Po sledam Pifagora*, Moscow, 1960–1961.

23. Within any quartet, notice that every *pair* (1,2), (3,4), (5,6), etc., is placed along a "knight's move" on the chessboard. Notice also that in each quartet there is one entry in each row, column, and diagonal.

24. Rows, columns, half-rows, and half-columns will always sum correctly. So will the two main diagonals and the four main bent rows (those that connect two corners of the square). According to that measure, the order doesn't matter; but other properties do depend on the order in which the 4×4 ingredients are joined together. However, all bent rows and diagonals (and most 2×2 blocks) will work for the particular configuration that appears in "More Magic Squares," http://www.pasles.com/magic.html.

25. Edward R. Hogan, "Robert Adrain: American Mathematician," *Historia Mathematica*, Vol. 4, 1977, p. 157.

26. Karen Hunger Parshall and David E. Rowe, *The Emergence of the American Mathematical Research Community, 1876–1900: J. J. Sylvester, Felix Klein, and E. H. Moore,* Providence: American Mathematical Society, 1994.

27. For example, in *The Mathematical Monthly* (1858) and in *The Analyst* (1879).

28. (The governor's other descendants include U.S. Senator John Kerry.) Among Professor Winthrop's undergraduate students was future president John Adams.

29. Franklin to John Winthrop, July 2, 1768, *Writings,* Vol. 5, pp. 136–142. He adds: "When I have the pleasure of seeing you, I will communicate it."

30. Charles Hutton, *A Mathematical and Philosophical Dictionary,* London, Printed by J. Davis. . . . , 1795–1796.

31. *A Catalogue of the Entire, Extensive and Very Rare Mathematical Library of Charles Hutton, L.L.D.* London: Wright and Murphy, Printers, 1816.

32. Hutton's course was used at the Royal Military Academy and at West Point. For the latter, see V. Frederick Rickey, "The First Century of Mathematics at West Point," in *History of Undergraduate Mathematics in America: Proceedings of a Conference Held at the United States Military Academy, West Point, NY, June 21–24, 2001,* ed. Amy Shell-Gellasch.

33. Charles Hutton, *A Philosophical and Mathematical Dictionary. A new edition, with numerous additions and improvements,* London: Printed for the author, 1815.

34. Obviously, Isaac Dalby was familiar with both diagrams long before the 1815 edition of Hutton.

35. Joseph Youle, *The arithmetical preceptor, or, A complete treatise of arithmetic, theoretical & practical. In six parts. To which is added A treatise on magic squares. . . .* Sheffield: Longman, Hurst, Rees, Orme & Brown, 1813; and Thomas Watson, *An useful compendium of many important and curious branches of science and general knowledge, digested, principally, in plain and instructive tables, to which are added some rational recreations in numbers, with easy and expeditious methods of constructing magic squares and speciments of some in the higher class,* London: Longman, 1812.

7 Circling the Square

. . . the good Clerk of Oxenford did show us a riddle
touching what hath been called the magic square.
Of a truth will I set before ye another that may seem
to be somewhat of a like kind. . . .
　　　—Chaucer's Franklin, in *The Canterbury Puzzles*[1]

*I*n letters to the English electricians, Franklin presented his Magic Circle of Circles (figure 7.1), ensuring that it would be published alongside other groundbreaking discoveries in the 1769 edition of *Experiments and Observations*.[2] With its twists and turns, it makes an imposing impression at first glance, even more so after its virtually supernatural symmetry is revealed.

Like the magic 16-square, Franklin's circle had seen its public unveiling two years earlier, in James Ferguson's *Tables and Tracts*. Ferguson called this circle "the first of its kind I ever heard of, or perhaps anyone [else] besides." Before introducing you to the intricacies of this fascinating, forbidding figure, I'd like to ask a question that hasn't been raised before: Just what on earth possessed Franklin to draw a magic *circle*?

Recall that mathematical designs are indispensable components of that vaguely scientific field known as intelligence testing. On first exposure, magic squares offer the quintessential opportunity to measure nonlinear thinking. (If you haven't seen a magic square

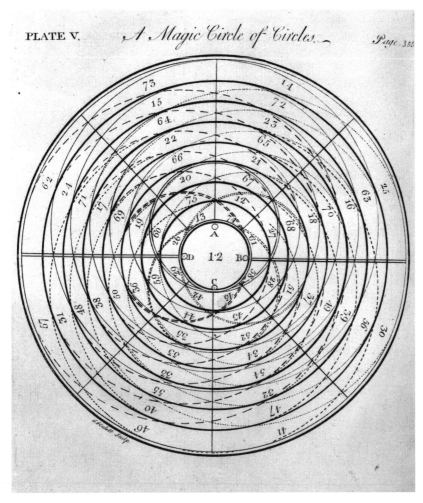

Fig. 7.1. *A Magic Circle of Circles*, plate V, p. 355, from Franklin's *Experiments and Observations on Electricity*, 1769. The Historical Society of Pennsylvania.

before, you must literally think in two directions in order to detect the pattern.) Open up a book of Mensa puzzles, or work through one of the faux intelligence tests that pollute the Misinformation Superhighway, and you are likely to run into a lo shu or one of its descendants. For not only is their creation a gymnastic for budding mathematicians, as Franklin and others have surmised, but their

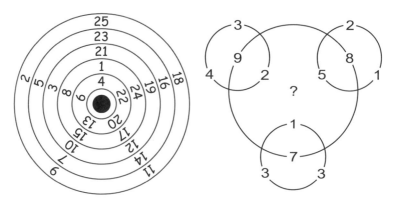

Fig. 7.2.

analysis is also a test of those same abilities. To create, to solve, to analyze: these are all separate talents measured by the mystical magic square.

In certain classic puzzle-books of the twentieth century, one finds a curious variation on this magic theme, intended to test one's penchant for creative thinking.

Figure 7.2, left is from *Mathematical Games*, a collection of puzzles originally published in Hungarian by the prolific writing team of Clara Lukács and Emma Tarján, since translated into English and available in various editions over the past forty years. Figure 7.2, right is from *100 Games of Logic* by Pierre Berloquin, a longtime contributor to the science supplement of *Le Monde*. These puzzles are no trivial footnotes, for each challenge is prominently displayed on the cover of the latest editions on my shelf. And both images descend from a family of historic pedigree which reached their culmination in the mathematical work of Benjamin Franklin.

The first puzzle is called *Move the Disks*. You are to construct disks of varying sizes out of paper or cardboard, embroider them with the numbers shown, and then fasten these together in the center to allow for each disk to be rotated independently of the others. Notice that the total around every circumference is the same; the challenge, then, is to rotate the disks until every radius likewise adds up to the same number.[3] The second item above is merely titled

Game 36, and the goal is to find the missing number.[4] Though they follow different outlines and come from different sources, what these two puzzles share is an affinity with another ancient artifact, the magic circle.

As far as I can tell from my own travels through the history of mathematics, no one has ever really traced the evolution of the magic circle. Certainly there has been no effort to connect traditional circles to their modern-day incarnation, nor to relate either the ancient or the modern variety to Franklin's own contribution. Personally, I have always been astounded that the term "magic circle" appears in the world's best-known *Autobiography* without context or explanation. Even in the dozens of editions now available, repackaged and annotated, not a one suggests that there was any such thing as a "magic circle" before Franklin, nor any link to a larger mathematical tradition. Yet here is a story in its own right, of an art form that has crossed many cultures over the course of three-quarters of a millennium.[5]

As with the history of magic squares, it is difficult to pinpoint just when magic circles were first conceptualized. Oftentimes, authors present work culled from prior sources that are no longer available. What is certain is that by the thirteenth century, magic circles had made their debut in China. In his 1275 work, the Chinese mathematician Yang Hui presented a half-dozen magic circles, two of which are reproduced in figure 7.3.[6] See if you can spot the rules and patterns in each one.

Do these look familiar? They ought to. These appear to be ancestors to the two given on the previous page. Start with the diagram on the left: Yang Hui intersects eight radii with four concentric circles, and places numbers at the intersections. The puzzle you saw before crosses five radii with five concentric bands instead, and when rearranged in its solved form it has the exact same sort of magic as Yang Hui's: sum along the radii or the circles to see it.

Meanwhile Yang Hui's other example, on the right, consists of four small circles whose centers are evenly arranged around a larger circle. Notice there are circles of three different sizes. For each, the sum of the central number with the four on its circumference

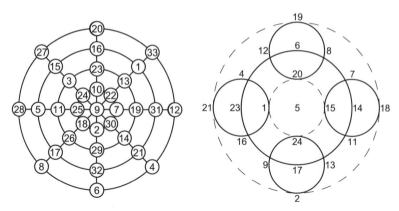

Fig. 7.3. Yang Hui, 1275.

always adds to 65. Compare Berloquin's puzzle (figure 7.2, right): three small circles whose centers are evenly arranged around a larger circle. And why stop at three or four small circles? Why not five, six, seven or eight? Figure 7.4 shows some later examples from seventeenth-century Japan.[7]

In each of these, the numbers lie on small circles whose centers fall evenly around a larger circle, itself sometimes invisible. In every case, the sums around the smaller circles are all equal. These three diagrams were called *magic wheels* by their creator, the mathematician Isomura.[8] If only by chance, they call to mind the ancient models of the heavens, wherein each planet followed a convoluted noncircular orbit, thereby evading the necessity of centering the universe at the sun. Such curves, called epicycloids, are generated by following a point on a small circle (epicycle) that rolls around a larger circle—the very essence of a magic wheel.[9] They are wheels within wheels, like the gears of a clock, meant to account for the apparently backward motion sometimes traced by heavenly bodies.

The other, concentric, style of magic circle prospered in Japan as well. Most typically, this included as many diameters as circles—which is to say, twice as many radii as circles—and whole numbers were placed at the intersection points of that circular grid.[10] An additional entry is placed in the center. So far, it sounds like Yang Hui's

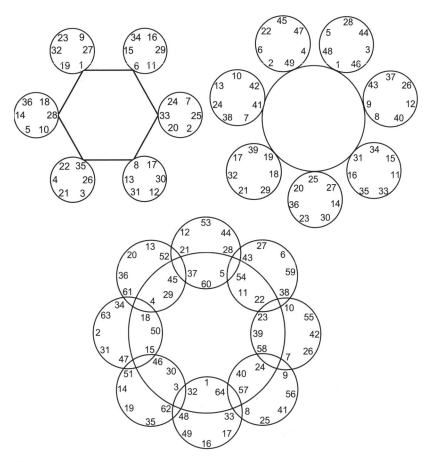

Fig. 7.4. Isomura's magic wheels, seventeenth century.

magic circle, but by the 1600s the Japanese mathematicians had made it standard practice to specifically put the number 1 in the middle, as in figure 7.5, in such a way that the sum around every circle is the same, and the sum along any diameter is the same. (The sum along each radius need not be the same.)[11]

One master of the magic circle in seventeenth-century Japan was Muramatsu Kudayu Mosei, who wrote numbers along 8 or 16 circles and as many diameters. Muramatsu was one of the Forty-Seven Ronin, the great samurai immortalized in kabuki theatre, cinema, song, and literature, who committed *seppuku* (ritual suicide) in the

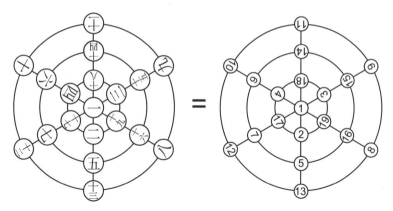

Fig. 7.5. Concentric-style magic circle.

time-honored fashion following an epic adventure in which their master's death was avenged. When not defending his leader's honor, he engaged in slightly less violent activities, such as computing π to eight decimal places and committing other mathematical atrocities.[12] More fortunate was Seki Kowa, born to the samurai class but destined for a less dramatic end. Seki has been called "Japan's Newton."[13] Like Muramatsu and Isomura, he was no dilettante; his mathematical range was impressive, and he invented the concept of a matrix *determinant* at least a decade before its supposed invention in the West.[14] Seki also invented a method of drawing magic circles of any size, so long as there were twice as many radii as circles. Magic circles after the style of Muramatsu and Seki are shown in figure 7.6. Add along diameters or circles to see the magic in each.[15]

While a surge of magic-circle-making took place in Japan between 1660 and the mid-1680s, there is no evidence of any immediate transmission to Europe—on paper, at least.[16] All of which brings us back to the Franklin circle that began this chapter. Since that graphic was fairly inscrutable with its confusion of dotted, dashed, and doubled circles, we begin instead with a scaled-down version in which those details have been removed, preserving only the essential elements (figure 7.7). To wit, I have erased every circle that does not share a common center with the diagram as a whole. Probably

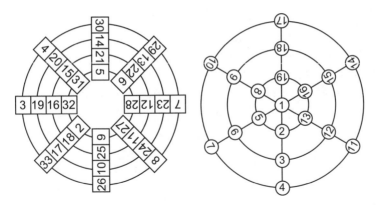

Fig. 7.6. More magic circles.

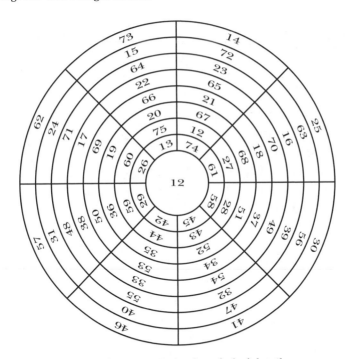

Fig. 7.7. Franklin's magic circle, denuded of details.

because it is so much easier to type and draw, this is how it appears in many books: essentially just a disk segmented into eight slices.[17] Yet while such a simplification makes plain the (incidental or intentional) affinity with Franklin's Asian predecessors—diagonals and even radii all sum equally, as do all concentric circles—it obscures his signature improvements.

For *improvement* is what Franklin sought, with his personal life and his inventions. There were almanacs before Franklin, but none like his. There were already stoves, eyeglasses, and magic squares, yet these were still to be improved. His Junto was aimed at the "mutual improvement" of its members. And so, with the magic circle, betterment was in order.

You are by now familiar with the more mundane aspects of such a design. Like its ancestors, Franklin's disk has the usual attributes: The sum along any one of the eight concentric circles is the same, 348; the sum along each diameter is the same, 708.[18]

The representation above lacks detail and thus appears to be no more impressive than its predecessors, but that misses the point entirely. The printed incarnation which graced the first page of this chapter is a far more accurate depiction. In fact, Franklin intended for the circle to be multichromatic,[19] but technological and economic considerations prevented its publication in that form. Through the magic of computer-aided design, I have restored his original intention in the color plates.

The colored circles, each centered at point *A*, *B*, *C*, or *D*, he called "excentric." The whole affair is a sort of combination of the two styles of magic circle. When blue, red, yellow, and green are removed, what remains is in the style of concentric magic circles. On the other hand if only the black circles are removed, we are left with an assortment of four-wheeled circles, five of them in all. Coincidentally, Franklin's term *excentric* dates to those same astronomical schemes already mentioned.[20]

Thus Franklin's particular genius was in producing a synthesis of both styles. Yet he might not have known of the earlier tradition. Perhaps he was only taking the logical next step in asking: if a square, then why not a circle? This would then be an example of *parallel evolution*, another way in which memes behave in accordance with biological change: the same alteration or innovation occurs independently in different places.

Here are the amazing properties of the magic circle, as Franklin understood them, slightly rephrased (see figure 7.8):[21]

- The sum along any radius, *including the center*, is 360, the number of degrees in a circle!
- The sum along any concentric black circle, if we include the central 12, is also 360.
- The sum along the upper or lower half of any concentric black circle—if we also include half the center ($6 = \frac{1}{2}$ of 12)—is 180, the number of degrees in a semicircle.

So far, so good: aside from the semicircle twist, which is novel, we have seen this animal before. But there's more (see figure 7.9):

- Each of the excentric circles, together with the central 12, sums to 360.
- Miraculously, these too can be partitioned along the particular diameters shown, paired with half of 12, and they sum to 180.[22]

Paraphrasing Franklin, James Ferguson writes:

> Observe that there is not one of the numbers but what belongs at least to two of the circular spaces; some to three, some to four, some to five: and yet they are all so placed, as never to break the required number 360 in any of the 28 circular spaces within the primitive circle.[23]

He also reveals a property not listed in Franklin's own descriptions: Draw a rectangle around four entries, and sum these with half the central 12 to obtain 180. There are fifty-six such patterns, two of which are delineated in figure 7.10.

Magic semicircles, excentric circles, and excentric semicircles are all innovations that began with Franklin. But just how did he effect this marvelous coincidence of pattern and number?

The secret, I believe, lies in a mysterious scrap of paper archived in the Franklin Papers of the American Philosophical Society. Unpublished during Franklin's lifetime—nor for nearly two centuries after it was authored—this sheet records an 8 × 8 magic square,

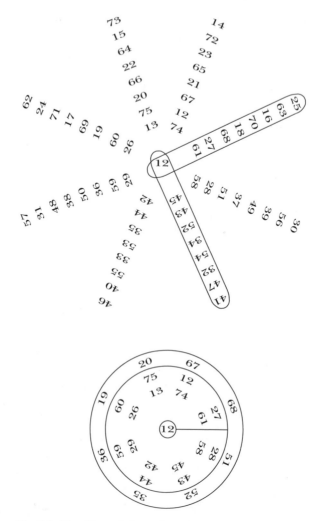

Fig. 7.8. Top: Two of the eight radial lines, each of which sums to 360. Bottom: One of the eight concentric circles, each of which sums to 360. Facing page: Two of the sixteen semicircles, each of which sums to 180.

similar in many ways to its more famous counterparts. Its specific origin unknown, this document includes just those 64 numbers, no other text and no context; adjacent documents in the collection apparently bear no connection with this magic square and thus

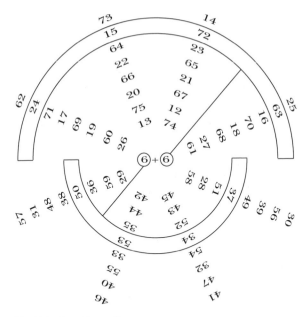

Fig. 7.8. (continued)

provide no clues as to its purpose.[24] Why wasn't the magic square shared with Franklin's many interested correspondents? Why bother to create a different 8-square, when that particular mountain had already been conquered? Perhaps, as I have argued in the pages of the *American Mathematical Monthly*, because this was a private document—a rough draft of the famed Magic Circle of Circles.

17	47	30	36	21	43	26	40
32	34	19	45	28	38	23	41
33	31	46	20	37	27	42	24
48	18	35	29	44	22	39	25
49	15	62	4	53	11	58	8
64	2	51	13	60	6	55	9
1	63	14	52	5	59	10	56
16	50	3	61	12	54	7	57

One of the astonishing properties built into this little gem is that you may slice off the first two or four or six columns, and

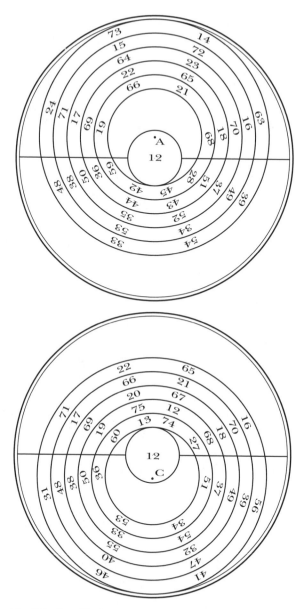

Fig. 7.9. Franklin's excentric circles, centered at
A, *B*, *C*, and *D*. Don't forget to include the middle 12
when you sum! Each can be split along the meridian
to produce half the magic sum.

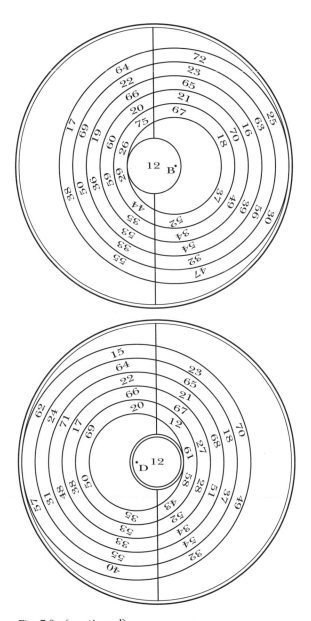

Fig. 7.9. (continued)

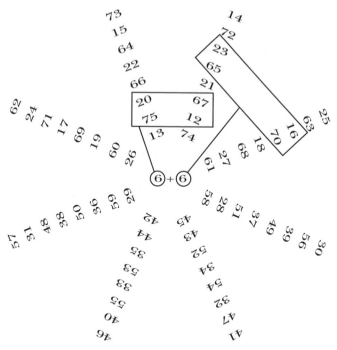

Fig. 7.10. Sums of 180.

transplant them to the right-hand side; you will find that all of the magical properties of Franklin's magic square still hold true: rows, columns, bent rows all sum to 260.[25] So, let's shift the first two columns to the end (see figure 7.11, left). Now, since the half-rows are half-magic (each sums to $\frac{1}{2}$ of 260, or 130), we can interchange those half-rows too. This is easily accomplished by flipping some 2×4s upside down (see figure 7.11, right). Recall that in Franklin's squares, every 2×4 rectangle sums magically; this is likely one of the additional properties he referred to enigmatically but left undisclosed in his letter to Peter Collinson, for lack of time or space (or out of a desire to keep a few secrets). The result is still a magic square with all of Franklin's usual miraculous mathematical qualities.

Now flip the sheet over and fan out the rows into radii, after the style of Muramatsu (figure 7.12).[26] Having spread the rows out

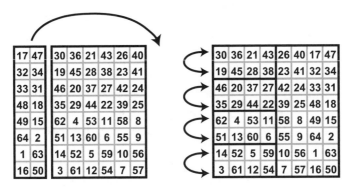

Fig. 7.11.

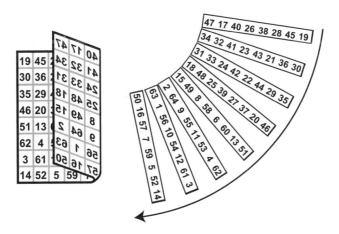

Fig. 7.12.

evenly, we obtain a magic circle whose sums are still 260. Since he is cleverly aiming for 360, he adds a central 1, as in the Asian magic circle tradition, then adds 11 to every entry (figure 7.13). Each circle and radius contains nine numbers now (including the central value), and so we have increased the magic sum by $1 + (9 \times 11) = 100$, that is, 260 increases to 360. And that explains why the entries of Franklin's magic circle are not 1 through 64 as you'd expect, but rather 12 through 75.[27]

In the penultimate step, we morphed the square into a circle by wrapping rectangular pieces around and turning them into the

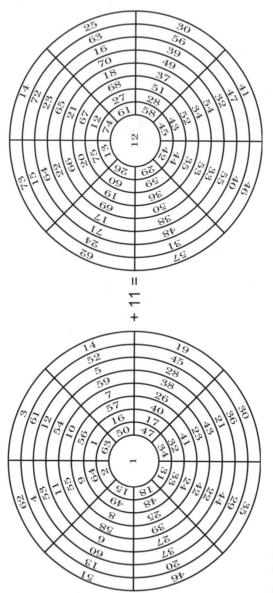

Fig. 7.13.

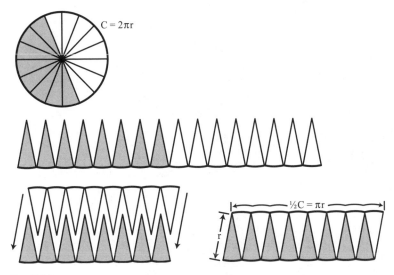

Fig. 7.14.

eight pizza slices of Franklin's magic circle. The same process performed in reverse is used in elementary mathematics courses to demonstrate the connection between area and circumference.[28] First you unravel the disk into near-triangles (figure 7.14). Next fit these pieces back together, like shark's teeth on two opposing mandibles, to form something resembling a parallelogram. The more we slice our pizza, the closer this shape becomes to a true rectangle, with dimensions $r \times \pi r$. The inescapable conclusion is that the area of a circle is $r \times \pi r$. Thus, in what seems to be a minor miracle, the same constant number "pi" appears in the formulas for both the area and circumference of a circle.

Transforming the square to a circle allows us to explain the magical patterns inherent in Franklin's circle. Rows become radii, while columns become circles, but what of the rest? Ten excentric circles are derived from bent rows—so now we know why Franklin favored those idiosyncratic bent rows over conventional straight diagonals—but the other ten excentric circles come from a little-heralded property built into most of Franklin's magic squares, the magic zigzag (figure 7.15).[29]

Fig. 7.15.

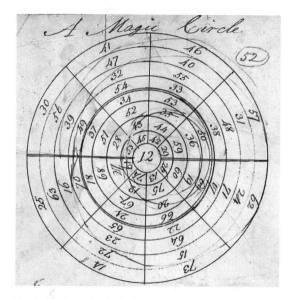

Fig. 7.16. Rough draft from a letter written by Franklin, probably in 1751. Some excentric circles are just barely visible. American Philosophical Society Library.

Although the *Autobiography* passage implies the existence of many magic circles, only this one specimen survives, unless we count obvious transformations like the 180° rotation seen in a rough draft (figure 7.16),[30] or the mirror reflection used in a later edition of Franklin's writings.

In earlier magic circles, it was quite common to have as many diameters as circles—which is to say, twice as many radii as circles.[31] Franklin follows a somewhat less commonly used tactic, so the number of radii is the same as the number of circles, in the style of "Move the Disks." The latter species is made by morphing a magic square, while the former is obtained by bending an oblong rectangle that is twice as long as it is wide.[32]

I do not believe that Franklin created his magic circles in a cultural vacuum, divorced entirely from prior tradition. Through some unknown source, perhaps a series of intermediaries, the eastern tradition must have reached him somehow. While magic squares followed a well-established route from East to West, conveyed multiple times with their source clearly acknowledged, the same is not true of the circle; the mathematician George Bruce Halsted says explicitly that the magic circle was "not rediscovered in Europe."[33] No one can say for certain whether Benjamin Franklin was aware of those predecessors when he drew his own circle, but the weight of circumstantial evidence is behind that hypothesis. His squares and circle are almost all based on 8 or 16, which are also the most common parameters used by his predecessors.[34] He places the smallest entry in the center. He follows the same general scheme, with concentric circles crossing diameters and constant sums along each radius and each annulus. He incorporates the other Asian style as well, for the Magic Circle of Circles includes five four-wheeled circles.[35] I mention only in passing that a thirteen-wheeled circle, sans numeration, appears in Franklin's papers (figure 7.17); this sketch and others like it served as the basis for bills and coinage, in particular for the first national cent.[36]

Even if he was completely unaware of a prior tradition of magic circles, and this is simply the memetic equivalent of parallel evolution, Franklin surely reached the pinnacle of this art.

While I have had to rely on conjecture in order to map the ancestry of Franklin's magic circle, its subsequent genealogy is somewhat easier to piece together. As with the copious commentary on Franklin's magic squares, the various authors who examined his circle are largely unaware of one another. Thus the literature is filled with repetitions and unwarranted claims of priority.

Fig. 7.17. Early draft of design used on the two-thirds dollar (1776), the Fugio cent (1787), and other denominations of bill and coin (reproduction shown). In some incarnations the wheels are adorned with the names of the thirteen colonies.

How exactly did Franklin's magic circle filter into the public consciousness? Its initial trajectory is closely tied to that of the famous magic squares of 8 and 16, for the circle was usually printed alongside one or both of these squares. From scattered sources, the following picture emerges.

Aside from that vague reference in the *Autobiography*—which might refer to some other magic circles no longer in evidence—our suspect first appears in Franklin's letters to the English electricians Peter Collinson and John Canton in the 1750s and 1760s. If there were further communiqués to the many other mathematicians of his acquaintance, these have not surfaced.

As we have seen, the circle finally made its public debut in James Ferguson's *Tables and Tracts*, a miscellany of mathematical and scientific items. Ferguson, a Scottish horologist/astronomer, had compiled material from a variety of authors, such as Thomas Simpson, a former editor of the *Ladies Diary*. Like Charles Hutton, Ferguson had risen to academic prominence and Royal Society Fellowship from modest beginnings (he was once a shepherd)[37]; his *Astronomy Explained Upon Sir Isaac Newton's Principles* was enormously popular. He explains:

[I] have lately got a magic square of squares and a magic circle of circles of a very extraordinary kind, from Dr. BENJAMIN FRANKLIN of

Philadelphia, with his leave to publish them. The magic square goes far beyond any thing of the kind I ever saw before; and the magic circle . . . is still more surprising.[38]

The following year, the magic circle and the square of 16 appeared in the *Gentleman's Magazine*, where Franklin's electrical discoveries had also been printed. In 1769, the fourth edition of the *Experiments* was published in London, and among its many updates were the magic circle and two magic squares. (But in each case, someone else was responsible for the editorial decision to include them; Franklin never published his own magical figures.) Various editions of Franklin's writings thereafter have printed the circle.[39] The magic circle next appeared alongside a map of the moon in the 1795 and 1815 editions of Hutton's great *Dictionary*, as well as in his later *Recreations*.

Continuing on the trail of the Franklin magic circle, I have found many amusing appearances. Like Winston Groom's *Forrest Gump*, or Woody Allen's *Zelig*, or like Franklin himself for that matter, it seems to be everywhere. One of the strangest early appearances was in an 1822 book called *Curiosities for the Ingenious, selected from the most authentic treasures of nature, science and art, biography, history, and general literature*. With a title like that, I knew my quarry must be near, and indeed the 16-square and the magic circle were both in attendance on the pages of this lively compendium of the miraculous. Most curious of all, neither square nor circle is given proper attribution. This is all the more surprising when one notices that another "curiosity" borrowed from Franklin's *Experiments*, concerning the properties of oil on water, *is* properly credited to him by name.[40] A sort of cross between *Scientific American* and the *National Enquirer*, this odd book ranges from the indisputably true to the laughably false: an account of a man who lived to the age of 180, a proof of the existence of unicorns, and experimental "evidence" that "savages" possess less physical strength than the "civilized" (the latter term referring to the English and the French). A strength-measuring device is illustrated on the same plate as the magic circle—itself stripped of most details, with not a single excentric circle outlined.

Possibly the weirdest curiosity in this little book is an alphabetic matrix, reminiscent of the famous SATOR square (chapter 2) but dwarfing it in size (figure 7.18).[41]

But this is hardly the end of the story for the Magic Circle of Circles. More significant than its simple reprinting—said appearances continued to proliferate throughout the 1800s—was the further evolution of the magic circle. Beginning in the early part of that century, a number of mathematicians saw fit to extol the virtues of Franklin's annular masterpiece, while simultaneously suggesting that it was deficient in certain particulars. In my studies of the literature, I have unearthed several such efforts, and they all share the same unwarranted criticism of Franklin: that he should have included spiral designs in his circle.

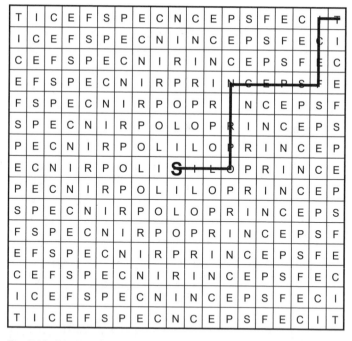

Fig. 7.18. Medieval search-a-word: an alphamagic "square" (really a 19 × 15 rectangle), this inscription is taken from the entrance to a Spanish monastery. The phrase "silo princeps fecit" can be traced in thousands of ways, one of which is shown here.

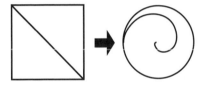

Fig. 7.19.

Why should this be so? In a conventional magic square, the diagonals are straight, not bent like Franklin's. When the square is transformed into a circular array, the diagonals become spirals instead of excentric circles (figure 7.19).[42] Franklin's detractors are basing their complaint on blind devotion to tradition.

Franklin's magic circle has been subject to fewer machinations and variations than his magic squares, perhaps only because it is harder to draw and reproduce. The first in-depth studies of this circle were undertaken in 1812–1813, by the Reverend Thomas Watson and Joseph Youle (see chapter 6).[43] The next, more significant commentary is due to Eugenius "Owen" Nulty.

An instructor in languages, sciences, and navigation, Nulty would be remembered years later as "a scholar of an old and rigorous type, and a man of much individuality and force."[44] He was a mathematics professor at Dickinson College (founded by Franklin compadre Benjamin Rush), and later worked as a "Calculator" for the U.S. Coast Survey (which had been created by Thomas Jefferson in 1807, and which is today part of the National Oceanic and Atmospheric Administration).[45] He received an honorary degree from the University of Pennsylvania.[46] As an American Philosophical Society member and an active local mathematician, he certainly had access to resources and oral tradition, and so he might have interacted with scientists who had been personally acquainted with the elderly Franklin. This may be what first interested Nulty in the magic circle.

Certainly he must have been considered a respected authority on the subject. When the APS considered whether to posthumously publish John Ewing's magic square, Nulty was called on to judge this matter. He gave the thumbs down, and his opinion carried the day. Ewing's square remained unpublished until the present book.[47]

Nulty completed two investigations of the magic circle, separated by an interval of sixteen years, with a break in between owing to a "nervous affliction" (which in those days might euphemize anything from actual physical complaints to moderate depression to a complete mental breakdown).[48] His prose is entirely without restraint. "The Magic Circle of Dr Franklin has been long admired, as embracing the most ingenious arrangement of numbers ever formed," he writes, for it "would seem to require a considerable familiarity with the powers of numbers" and was not found "either by conjecture or trial." Later he reiterates: "it has been considered in Europe as the most ingenious arrangement of numbers ever imagined."[49] Nevertheless, he alleges, Franklin's circle is *imperfect*.[50] Nulty's improvement, replete with ten spirals each adding to 360, is shown in figure 7.20,

Fig. 7.20. Top left: An early imitator's "improvement" of the Franklin circle. Bottom: The same with only clockwise spirals highlighted. See the color plates for more detail. Top right: A later attempt, with one spiral and one excentric circle visible. The spiralists recall the strange and misdirected fundamentalism of some critics we met in the previous chapter.

upper left.[51] He calls his creation a *cyclovolute*: literally, a circle of spirals, or volutes.[52]

While Nulty copied the same confusing dotted and dashed circles into his cyclovolute, he did have the inspiration to use colors to distinguish spirals. These are Archimedean spirals, to be distinguished from the logarithmic spirals that grace the chambered nautilus shell. The spirals in Nulty's cyclovolute are clockwise or counterclockwise depending on whether they were sired by forward-slash or back-slash diagonals in the original square.[53] The overall effect resembles a swirled lollipop (see the color plates).

Watson and Nulty did not simply replace circles with spirals, but rather found a way to include both. But to claim they have thereby bested Franklin is like repainting your Chevy and claiming to have invented the automobile.

And so it goes through other authors. The most prominent of these is F.A.P. Barnard, whose spiral-laden circle appears in figure 7.20, upper right.[54] This is not a man who would be deterred by the apparently useless aspect of the magic square, for he once expounded in great detail on the many seemingly insignificant discoveries which lead to great scientific advances, explaining:

> that there is no new truth whatever, no matter how wide a space may seem, in the hour of its discovery, to divide it from any connection with the material interests of man, which carries not within it the latent seeds of a utility. . . .[55]

Barnard, like Nulty, worked for the Coast Survey. Coincidentally, both men served under the leadership of superintendent Alexander Dallas Bache, great-grandson of Benjamin Franklin.[56] Barnard was also Chancellor of the University of Mississippi until the Civil War broke out, he was President of the American Association for the Advancement of Science, and later President of Columbia College. Today, Barnard College of Columbia University bears his name.

In his monograph on magic squares and cubes, Barnard refers to "the marvelous character of the arrangement" and "an ingenious geometrical artifice" of the Magic Circle of Circles.[57] Yet like Watson, Nulty, and others, he seeks to improve Franklin's magic circle by adding spiral magic sums. It is difficult to imagine why anyone would feel intuitively that spirals are aesthetically superior to circles; indeed, the Greek notion of circular perfection is part of the reason that the erroneous idea of circular orbits persisted for so long in astronomy. So why did Franklin's critics find it necessary to further "improve" his work with spirals? Along with the carping Momuses of chapter 6, these critics were motivated by a blind devotion to the past, no less than the overzealous followers of Aristotle and Ptolemy. We will rebut their poorly aimed critiques in the next chapter, and allow Franklin himself to set the record "straight." Meanwhile we can express only wonderment and awe at the sheer number of symmetries packed into Franklin's squares and circle.[58]

Notes

1. Henry Ernest Dudeney, *The Canterbury Puzzles and Other Curious Problems* (4th edition), New York: Dover Publications, 1958, p. 45.

2. The earliest letter to mention the circle was written to Peter Collinson around 1752. It was published, possibly in edited form, in the 1769 edition of *Experiments and Observations*. (The manuscript is apparently no longer extant; however, the American Philosophical Society Library holds a draft copy.)

3. Clara Lukács and Emma Tarján, *Mathematical Games* (originally *Vidam Matematika*), 1963; English translation, 1968; Barnes & Noble reprint, 1996, pp. 136–137. I have taken a slight liberty and scrambled the disks; in its original form, the problem is posed as its answer key, and the reader is advised to scramble it and challenge a friend. For me, that spoils the fun, and also camouflages the fact that—unknown to the authors, apparently—there are actually three different correct answers. (It should be noted that the authors' answer has ten additional spiral-shaped magic sums, whereas the others possess only six each; however, that's not mentioned as part of the challenge.) *Solution*: It stands to reason that we lose nothing, and gain plenty, by assuming that the outermost disk is already in its proper orientation. From our previous discussions of magic squares, you should be able to see rather quickly that, since there are as many radii as there are circular bands, the sum we are aiming for must be 65. Thus, you could simply try out all 25 possible combinations of the next two disks, in each case calculating the vertical radius sum. If it falls short of 65, determine whether that shortfall can be compensated by the two remaining disks, and then check the other radii to see if they, too, sum to 65.

A quicker and more satisfying intuitive approach is as follows: One must not gather too many small numbers (nor too many large numbers) in the same radius. For example, if 3, 4, and 5 appeared in the same line then the other two values in that radius would have to fill out to 65 by contributing 53. That's impossible with our values (no entry greater than 25). Intuitively, then, why not spread out the smallest (and largest) entries as much as possible, placing 1,2,3,4,5 in successive radii? Try it clockwise and counterclockwise, and one of these works out to a correct solution. *Final answer:* You should align the disks so that one radius reads 4,15,21,7,18; or 6,24,12,5,18; or 4,8,12,16,25.

4. Pierre Berloquin's *100 Games of Logic*, 1977; Barnes & Noble reprint, 1995. *Solution:* For each circle, you can add up the numbers around the circumference to get the number in the center. Thus, the answer is 24.

5. The topic has even appeared in the grade school curriculum. Elizabeth Stadtlander, "Arithmetic Theories, A State Course of Study, and Textbooks," *The Elementary School Journal*, Vol. 41, No. 6, 1941, pp. 438–453.

6. The first diagram is based on Joseph Needham, *Science and Civilisation in China, Vol. 3: Mathematics and the Sciences of the Heavens and the Earth*, New York: Cambridge University Press, 1959. The second image is patterned after George Gheverghese Joseph, *Crest of the Peacock*, new edition, Princeton University Press, 2000. Apparently no secondary source includes more than one of these images, which obscures the fact that Yang Hui knew more than one distinct style of magic circle.

7. David Eugene Smith and Yoshio Mikami, *A History of Japanese Mathematics*, Chicago: Open Court, 1914.

8. A somewhat different variation on the theme appears in China, probably sixteenth century:

```
      5  68          3  70          1  72
   32        41   34        39   36        37
   42        31   40        33   38        35
      67  6          69  4          71  2


      18  55         16  57         14  59
   19        54   21        52   23        50
   53        20   51        22   49        24
      56  17         58  15         60  13


      11  62          9  64          7  66
   26        47   28        45   30        43
   48        25   46        27   44        29
      61  12         63  10         65  8
```

Here the sum around every circle is the same (292). Row sums are all equal (219), and column sums alternate between 220 and 218. After Frank J. Swetz,

"Seeking Relevance? Try the History of Mathematics," *Mathematics Teacher*, Vol. 77, 1984, pp. 54–62. (I have made a minor numerical correction.) Swetz does not specify the exact source, but places it in the Ming period (1368–1644). I suspect that that this diagram appears in Ch'êng Tai-Wei, *Suan-fa T'ung-tsung* (*A Systematized Treatise on Arithmetic*), 1593, based on the description of the latter text in Yoshio Mikami, *The Development of Mathematics in China and Japan,* New York: Chelsea, reprint of 1913 edition, pp. 110–112.

9. Finding the equations of an epicycloid is a standard calculus exercise which can be found (for example) in Sherman K. Stein, *Calculus and Analytic Geometry*, 4th edition, pp. 461 and S-132.

10. Seventeenth-century Japanese mathematicians did consider the case where radii and circles are equally numerous; for example Muramatsu and Nozawa Teichō (Smith & Mikami, p. 80).

11. The information on summing properties is taken from Smith & Mikami, 1914, and Vera Sanford, "Magic Circles," *Mathematics Teacher*, Vol. 16, 1923, pp. 348–349. The example I gave is in that same style, but as far as I know not from that era; I have invented it for the purposes of this discussion.

12. George Bruce Halsted, "Pi in Asia," *American Mathematical Monthly*, Vol. 15, 1908, p. 84. Halsted refers to Muramatsu by the variant name Shigekiyo Matsumura, but since he names the latter as the author of *Sanso* (1663), they must be one and the same person. Halsted wrote the preface to *The Development of Mathematics in China and Japan*, and he is among the few mathematicians specifically acknowledged by its author. Some of the Halsted-Mikami correspondence is preserved at the Archives of American Mathematics, Center for American History, The University of Texas at Austin.

13. However, Smith and Mikami demur: "He may be compared with Christian Wolf rather than Leibniz, and with Barrow rather than Newton" (p. 127).

14. Credit is ordinarily given to Leibniz. The determinant tells, among other things, whether a square matrix has an inverse, something like the reciprocal of a number. (Almost every number has a reciprocal, the exception being zero; but many square matrices lack an inverse, so this is a deeper concept.)

15. The left-hand figure does in fact have constant radii as well. This is a feature common to the circles of Yang Hui, Muramatsu, and Franklin.

16. These dates are compiled from Smith and Mikami, Needham, and Joseph, based on works which included concentric circles and wheels.

17. For example, the simplified circle appears in an 1822 book of "curiosities" described later in the present chapter. Likewise, in the manuscript version reproduced photographically on p. 399 of Paul Leicester Ford's "The Many-Sided Franklin: Franklin's Schooling and Self-Culture, His Services to Education, His Library" (*The Century*, Vol. 57, No. 3, 1899, pp. 395–414), the non-concentric circles have all been removed (possibly through no fault of the author, for these were drawn in a different color and may not have survived photographic reproduction). Such oversimplification is not confined

to the nineteenth century. More recently, in 2002, one popular mathematics book included a version so simplified that even the basic properties fail to work; by leaving out the central entry and miscopying another part, the author ensures that Franklin's intended sum of 360 does not appear even once (and it varies instead between 347 and 348).

18. One subtle point should be addressed here. In Franklin's circle, as with his squares, I have tried to follow his own terminology even when it conflicts with modern usage. Thus we use the term "bent row" even though these look more like diagonals than rows. Likewise one speaks of summing along a circle, even though technically these entries lie in an *annulus*, which is to say *between* two circles. The sum along any one of the eight concentric annuli is 348. (The terms *circle*, *annulus*, *band*, and *ring* are often used interchangeably, even though their precise definitions do not coincide.)

19. The color version is described in a letter to Collinson (published in *Experiments and Observations*, 1769, pp. 354–356) and drawn in a letter to John Canton (May 29, 1765).

20. In said theories it is sometimes even spelled in that unusual fashion. In pre-Copernican astronomy, an excentric (eccentric) orbit is centered at a point different from the Ptolemaic center of the universe (the Earth).

21. These are taken from multiple sources. He explained the details a little bit differently each time: in one letter to Collinson, another to Canton, probably by proxy in Ferguson's book, and in one later instance which survives only in translation. Taken together, these properties prove beyond doubt that he built his circle from a magic square. For example note the little-known block property in Ferguson, a series of fifty-six rectangular patterns described in figure 7.10.

22. Technically these are not really halved, because the diameters shown pass through the central 12 and not through point *A*, *B*, *C*, or *D*.

23. Ferguson, *Tables and Tracts*.

24. Franklin Papers of the APS LXIX, 104. The *Papers*, Vol. IV, 1961, p. 395, footnote 7 attributes the square to Franklin.

25. Of all of the evidence that this property is intentionally built into his magic square, I find no clearer reason than the fact that something quite similar is true of his famous 8-square as well: instead of stripping off double columns, one slices off the double rows of that square.

26. The flip-over step is not really necessary. If you don't flip the sheet over, you get a mirror image of the Franklin circle, instead—but this may be how he originally envisioned it anyway. (I discovered recently that in at least two editions, Franklin allowed it to be published in this reflected form.)

27. He might have first tried an alternate strategy to obtain his 360: add 12, not 11, to every entry, and place a 4 in the center instead of a 12. Indeed there is evidence Franklin considered this alternate move, since he mistakenly wrote that the entries were "13 to 76" then crossed that out. In the corrected versions, he points out that his circle contains the numbers from 12 to 75, cagily sidestepping the fact that there is one minor repetition: the 12

appears twice. (Magic squares with an entry or two that are repeated twice are examined in Pasles, 2004.)

Since I first proposed this construction, mathematician-journalist Barry Cipra has pointed out that the same magic circle can also be obtained from the more famous magic 8-square. (Personal communication, April 12, 2001.) However, I believe that his construction is less likely—though still a possibility. It is even possible that Franklin built his magic circle from some other, undiscovered, square.

28. For example, such a diagram is found in the Keedy/Bittinger series of textbooks.

29. See Pasles, 2001 and 2003. Zigzags are built into both of the 8-squares we have seen so far, although in the case of the more famous 8-square you must first turn the diagram sideways for the zigzags to work. The same is true of the 16-square, and of another large magic square we will see in the next chapter. Suffice to say, this appears to be an intentional feature of Franklin's magic squares.

There are many other ways in which the two 8-squares appear to be sideways versions of one another. Most obviously, as you can check, the last two rows of the first square are identical to the first two columns of the other square.

30. The same image was printed in Ford's "The Many-Sided Franklin" (see note 17), but with excentric circles removed from both the image and the description—an unfortunate deletion of the most innovative aspect of Franklin's circle.

31. Halsted, 1908; Smith and Mikami, 1914; Sanford, 1923; Sanford, *A Short History of Mathematics*, New York: Houghton Mifflin, 1930; Sr. M. Nicholas, "Magic Circles," *American Mathematical Monthly*, Vol. 62, 1955, p. 696.

32. In that case one uses something like a magic rectangle, but a little bit more complicated.

33. Halsted, 1908.

34. This may however mark nothing more than the mutual recognition that it is easier to divide a circle into eight equal parts than, say, seven or eleven parts (as anyone who has ever sliced a pizza for exactly seven guests can attest): simply bisect the disk repeatedly. (Interestingly, Muramatsu's value of pi was produced by this same process of continued bisection.)

35. Each of these reasons alone is unremarkable, but together they have some force. As Edgar Allen Poe famously wrote in *The Mystery of Marie Rogêt*, independent reasons when combined constitute "proof not *added* to proof, but *multiplied.*"

36. (I will resist the urge to comment on the "one" in the center.) This diagram is based on a sketch that appears in the *Papers*, Vol. 22, 1982, plate facing p. 358. While the provenance of that particular drawing is not certain, the editors determine based on other evidence "the conclusion is hard to escape that Franklin was the designer," pp. 357–358. (Their assertion is based on arguments made by Eric Newman in *The Early Paper Money of America*, 1967.)

In the final version that appears on paper money and coins, the largest circle has been eliminated.

The draft looks very similar to the twelve-wheeled circle on p. 343 of *Divers Ouvrages*, the book Logan shared with Franklin on that fateful day (see chapter 5), though neither of these is a (numbered) magic circle.

37. The same could be said of Simpson, who was a weaver by trade.

38. *Tables and Tracts*, p. 309.

39. In English, French, and German; but not in the Italian edition, which was much abbreviated.

40. Anonymous, *Curiosities for the ingenious; selected from the most authentic treasures of nature, science and art, biography, history, and general literature, Second edition, with Improvements,* London: Printed for Thomas Boys, Ludgate Hill, 1822. A later 1825 American printing of this title may constitute one of the earliest stateside appearances of the magic circle (not counting imported books, which were still in good supply).

41. More information on this monastery inscription can be found at http://www.grao.net/ppueblos/santiapravia/santipra.htm. There it is claimed that the puzzle admits more than 250 solutions, whereas *Curious* claims precisely 270. In fact, by applying a standard theorem on such "lattice paths," it can be shown that there are tens of thousands of possible solutions. The precise mathematical expression for this number is $4 \times 16!/(7!\ 9!) = 45{,}760$. Kindred conundrums confront the pilgrims in Dudeney's *Canterbury Puzzles,* albeit diamond-shaped (p. 56) and triangular (p. 64, a variation on the classic "abracadabra" incantation triangle). The number of solutions differs from the Spanish inscription because these diagrams are not rectangular. Diamond-shaped arrays of the same type comprise eleven of Berloquin's *100 Numerical Games*.

42. See for instance John Calvin McCoy, "The Anatomy of Magic Squares #4", *Scripta Mathematica,* Vol. 4, 1939, p. 177. McCoy represents his circle as a hexagon, but the principle is the same.

43. Youle's magic circle of circles is a thinly veiled corruption of Franklin's own.

44. Edward Potts Cheyney, "On the Life and Works of Henry Charles Lea," *Proceedings of the APS,* Vol. 50, No. 202, 1911, pp. v–xli.

45. U.S. Census records for 1840 and 1860. Also Charles Coleman Sellers, *Dickinson College: A History*, Middletown, Conn.: Wesleyan University Press, 1973; and James Henry Morgan, *Dickinson College: The History of One Hundred And Fifty Years, 1783–1933*, Carlisle, Penn.: Dickinson College, 1933.

46. http://www.upenn.edu/commencement/hist/hondegalph.html

47. Letter of E. Nulty to R. M. Patterson, December 11, 1844. American Philosophical Society Library. Nulty's opinion is a bit disingenuous; he offers to correct Ewing's defects, if the APS will then publish this improvement instead—though he promises to mention Ewing's contribution, of course. (He certainly has his own interests at heart.) Also I find it suggestive that Nulty waited until five days after his own paper was allowed to be read at the APS before tendering a critical opinion of this celebrated former member.

48. Letter of E. Nulty to R. M. Patterson, November 26, 1844. American Philosophical Society Library.

49. E. Nulty, "A Remarkable Arrangement of Numbers, Constituting a Magic Cyclovolute." Read before the APS on June 27, 1834. *Transactions of the American Philosophical Society, New Series,* Vol. 5, 1837, pp. 205–208.

50. E. Nulty, "Supplementary Note on the Construction and different Forms of the Magic Cyclovolute." Read December 6, 1844. *Transactions of the American Philosophical Society, New Series,* Vol. 10, 1853, pp. 17–25.

51. Nulty, 1834. The colored spirals appear in some printed copies of the *Transactions*; in others, and in the electronic version available in many libraries, they are confusingly outlined in black and white.

52. Mathematicians love to name things. Various plane curves are described as evolutes and involutes; in biology spiral seashells are classified as volutes or *Volutidae*.

53. If you worked out "Move the Disks" at the beginning of this chapter, your solution will have either five or ten such spirals.

54. F.A.P. Barnard, *Theory of Magic Squares and Cubes*, Memoirs of the National Academy of Sciences, Vol. 4, Part 1, Washington D.C., 1888, pp. 209–270, which covered magic squares, cubes, circles, spheres, and cylinders. An abbreviated version of this material also appears as a disproportionately large entry in *Johnson's Cyclopedia*, of which Barnard was not coincidentally editor-in-chief (together with Arnold Guyot).

55. *Letter to the honorable, the Board of trustees of the University of Mississippi*, 1858.

56. Nulty, Barnard, and Bache were all members of the APS.

57. On the latter score his analysis is misguided: he claims that the whole structure works only because Franklin made the innermost circle three times as wide as the distance between that circle and the next one, but that measurement is incorrect (and irrelevant).

58. Some modern studies of the magic circle concept are Harry A. Sayles, "Magic Circles and Spheres," *Monist*, Vol. 20, 1910, pp. 454–471; Royal V. Heath, "A Magic Circle," *Scripta Mathematica*, Vol. 3, 1935, p. 340; and S. W. McInnis, "Magic Circles," *American Mathematical Monthly*, Vol. 60, 1953, pp. 347–351. A more recent proliferation of articles too populous to enumerate here includes little that is truly new.

8   Newly Unearthed Discoveries

His most famous square was a king-size brainteaser
that did not sum correctly at the diagonals, unless
the diagonals were bent like boomerangs. Now
that's flair, plus he dodged electrocution by kite.
   —Steve Martin, *The Pleasure of My Company:*
    A Novel (2003)

*W*e left our chronological narrative in order to examine
Franklin's magical meanderings. It is time to do some catching up.

Franklin's single-minded attention to investigations in natural
philosophy, enabled by his business success, would not last. Public
duties called, and these obligations would occupy the greater part
of his time for the remainder of his life. It is telling that he once advised a colleague: "let not your love of philosophical amusements
have more than its due weight with you. Had Newton been pilot of
but a single common ship, the finest of his discoveries would
scarcely have excused or atoned for his abandoning the helm one
hour in time of danger; how much less if she had carried the fate of
the commonwealth."[1] If there was little time for electrical experiments, there was certainly even less opportunity for arithmetical
amusements.

His public service, long in evidence, now increased apace. Following election to the Pennsylvania Assembly, he cofounded the

first public hospital in the Americas. He initiated the Philadelphia Contributionship to insure homes against fire. Even his commercial concerns were entwined with service to humanity. Both the *Gazette* and the almanac dispensed instructions on the construction of a lightning rod; he never patented this or any other invention. He became a Deputy Postmaster General for America and established the colony's first militia, and all of this occurred between 1751 and 1755.

Meanwhile his scientific endeavors earned copious accolades. The Royal Society awarded him its coveted Copley medal, and he received honorary master's degrees from Harvard, Yale, and the College of William and Mary. In 1756, he was elected an F.R.S., a Fellow of the Royal Society. At the time he was on the western frontier, where he commanded several hundred militiamen. After his return, he returned to his seat in the House, but early in 1757 he was appointed by the Assembly "to go home to England" on urgent business. Cost of war with the French and their Native American allies, they argued, required that all landowners in Pennsylvania must be taxed—even the Proprietors. These sons of William Penn enjoyed an exemption on their estate which, it was hoped, would be removed by the King and Parliament at the request of Philadelphia's favorite adopted son. Following bureaucratic delays, Franklin finally sailed for England in June, accompanied by son William but not his wife and daughter. En route, Poor Richard composed his final almanac.

This second visit to London would last five years. During this time, his wife Deborah was required to keep careful accounts and send regular financial reports to London. Franklin accomplished his mission, reaching a negotiated settlement with the Penns. He traveled to the north of England, where he met with cousins and found his ancestors' graves; and to Scotland, where he was received by the Masonic lodge. He received honorary doctorates from St. Andrews and Oxford. (A third doctorate, from the University of Pennsylvania, would not be awarded until 186 years after his death.) In 1762, just prior to his departure for Philadelphia, he found time to contrive the "glass armonica." This musical instrument consisted of glasses of gradually decreasing size, each one turned on its side and connected to a rotating rod. Like Franklin's concentric magic circles, these varying circumferences were marked by color: red for C, orange for D, and so on, with the "black keys" painted white. As the rod turned, the armonica player touched wet fingers to the appropriate glasses to call up eerie, beautiful tones that were said to drive mad both listener and performer. Legend aside, the instrument's popularity grew throughout the late 1700s, attracting the attention of popular audiences and the great composers. Its tonal accuracy was guaranteed by the precise specifications Franklin prescribed for making each glass. This was a feat requiring equal parts engineering skill and musical inspiration.[2]

Franklin's homecoming was short lived. In 1764, in short order he was elevated to Speaker, then lost reelection to the Assembly, and finally was sent back to London in official capacity, this time to fight the Stamp Act. (As a printer, he would be among the hardest hit by this tax.) These sojourns to the homeland allowed Franklin to cultivate in person several friendships that had been conducted exclusively by post. Among the scientific correspondents now within reach were Collinson and John Canton, both electricians with at least a passing interest in magic.

If you comb the literature, both historical and mathematical, from Franklin's time all the way through the end of the twentieth century, you will get the impression that only two of his magic squares were ever printed. The only exceptions were the alternate 8-square described in the previous chapter, and two smaller examples we will see in a moment. But these three appeared just one time, in a historical reference published in the 1960s, and as far as I can tell they went completely unnoticed by mathematicians and puzzlists. Articles on Franklin's squares continued apace, but these continued to repeat the same two examples over and over again, giving the impression that Franklin's numerical abilities were limited to these creations alone. What's worse is that the two magic squares are both manifestations of a single method, with the larger of these essentially a scaled-up version of the other. To see this, follow the placement of the values 1,2,3,4, and so on as they wind through the 8-square (chapter 5), then do the same with the

16-square. Can you see how the pattern generalizes to higher orders? You can test your theory by considering how one could use the same pattern to fill a 24 × 24 or 32 × 32 grid.[3]

Of course, this sheds no light on how Franklin came up with his method in the first place. My ability to play a Bach fugue on the piano is hardly equal to the composer's accomplishment. The metaphor can be carried further. If only one of Johann Sebastian Bach's works had survived, would we conclude that he had written that single piece and no other? If you came face to face with the *Mona Lisa*, and had never heard of da Vinci, wouldn't you assume that it was part of a larger corpus; that there must have been, at the very least, some lesser works that built up to this masterpiece? Without more evidence, though, a skeptic might conclude that Franklin had simply memorized his square—with a little practice it is very easy to recreate from memory—perhaps having learned it from someone else. But I always believed that there were more of Benjamin Franklin's numbers out there, and that it was only a matter of looking hard enough. Some authors have speculated that Franklin drew many different examples, but they seem to be operating on faith and not evidence. It's true, his autobiography refers to the making of magic *squares* and *circles*, plural nouns, but that description is not as clear as might at first appear. After all, Franklin titled his published diagrams as "a magic square of *squares*" and "a magic circle of *circles*," so it is conceivable that he referred to a single example. So where is that evidence?

The answer lies in Franklin's correspondence with fellow Fellow and Copley medalist John Canton, the first English physicist to confirm Franklin's results in electricity. A contributor to the *Ladies Diary* and the *Gentleman's Magazine*, Canton served with Franklin on the committee to evaluate the effectiveness of the "pointed conductors" on buildings. There was some concern that the pointed rod would not only redirect the path of lightning, but would actually increase the chance of a strike. Some rods were made to terminate in a ball instead, to reduce the hypothetical risk. The controversy eventually took on political implications, with differences of opinion varying by region. In later years Franklin would even make light of the debate in a metaphor: "The King's changing his pointed conductors for blunt ones is therefore a matter of small importance to me. If I had a wish about it, it would be that he had rejected them altogether as ineffectual, for it is only since he thought himself and family safe from the thunder of heaven, that he dared to use his own thunder in destroying his innocent subjects."[4]

In 1765, in the midst of the Stamp Act dispute, Franklin wrote a letter to Canton describing his magic circle. The letter makes clear that this was not the only time they had broached the subject. However, any earlier discussions must have taken place in person, or else were contained in one of those many letters now missing. Appended to this missive was not only the circle, but several magic squares of a type very different from his known works. One of these was identical to Dürer's 4-square (chapter 2). Was this a coincidence, or had Franklin seen *Melencolia I*? There are nearly 900 different 4-squares with straight diagonals, or more than 7,000 if we allow for rotations and reflections. (Recall that every magic square can be flipped or rotated to get seven more examples.) On the other hand, Dürer's square has some additional properties that a numerist like Franklin would surely have appreciated, so it may be a case of independent discovery. Of all of the magic squares that can be traced back to Franklin, this is the only one that lacks the proper bent-row sums.

The same letter also included a 6 × 6 magic square that is entirely original, one that displays very beautiful symmetries (figure 8.1). Though the *Papers of Benjamin Franklin* printed this square in 1968, I was not personally aware of it until I began to study Franklin in 1999. No one else seems to have noticed it either; the literature through the end of the twentieth century continued to focus on the 8-square and 16-square alone. All the better for us, for now we can have the pleasure of analyzing a Franklin magic square that has not been investigated before.

First, let's see what makes it "magic." By now you will be unsurprised to hear that every row and column sums to the same value, in this case 111. It contains some "bent rows" too; the accompanying text prescribes that "The diagonals are to be reckon'd by halves,

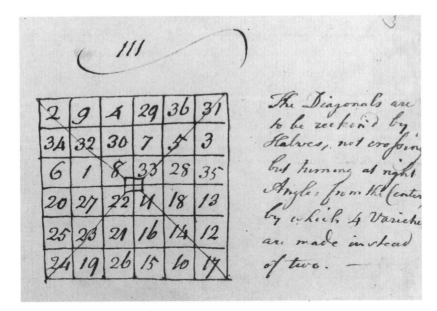

Fig. 8.1. "The Diagonals are to be reckon'd by Halves, not crossing but turning at right Angles from the Center, by which 4 Varieties are made instead of two." *Canton Papers of the Royal Society*, © The Royal Society.

not crossing but turning at right angles from the center, by which 4 varieties are made instead of two." These and many other inherent patterns are displayed in figure 8.2, every one of them summing to 111. Try to find more patterns on your own. The goal is not simply to locate six numbers that add up to 111; that can be done in just as many ways with any matrix that includes the numbers from 1 to 36. Rather, you can find that sum using patterns that are pleasing to the eye. (Some of these configurations can be found by using the fact that any three complement pairs will add up to the magic sum. It helps to observe the strange symmetry in this square: draw a line segment between each pair of complements: 1 and 36, 2 and 35, and so on; you should end up with a squished asterisk occupying the northern hemisquare, and another asterisk in the lower half.)

The 6-square is wonderful because it illustrates a method that differs radically from Franklin's other techniques. What's more, it is almost identical to a method that is generally credited to a

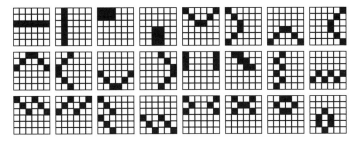

Fig. 8.2.

twentieth-century discovery by an English blueblood named Ralph Strachey.

Nearly every reference work on magic squares recites Strachey's method, though often without attribution. All of these diverse accounts appear to derive from the same source, the 1919 edition of *Mathematical Recreations and Essays* by W. W. Rouse Ball, the esteemed historian of mathematics and a successor to the recreational tradition of Ozanam and Hutton.[5] Ball was such a fan of recreational matrices that he used an alphamagic square (in the style of the monastery inscription, chapter 7) in his personal bookplate:

W	W	R	B
W	R	B	A
R	B	A	L
B	A	L	L

But who was Ralph Strachey? For hundreds of years, the Stracheys have produced accomplished authors, scholars, diplomats, and other notables. Within their ranks were intimate acquaintances of John Donne, John Locke, George Eliot, and Theodore Roosevelt. Charles Darwin's cousin Sir Francis Galton, the pioneering geneticist and statistician—and less admirably, a founder of eugenics— characterized them as "An old family, small in numbers, but of a marked and persistent type. Among its characteristics are an active interest in public matters, and an administrative aptitude."[6] Accordingly, Ralph Strachey's mathematical talents led him to become Chief Engineer of the East Indian Railway. As it happens, he also produced magic squares and stars. Toward the end of his life, he formulated the procedure we will now examine.

To illustrate the Franklin-Strachey methods, we start by fastening together four identical copies of the 3×3 magic square into a 6×6 grid (figure 8.3, left). Franklin begins with the lo shu matrix, Strachey with its 180° rotation, but their methods will work equally well with any one of the eight 3×3 matrices described at the end of chapter 1. The resulting 6×6 matrix is certainly magic, but it's not very interesting because every element repeats four times. To correct that deficiency, transform the matrix as follows:

- Leave the northwest panel alone, so we keep the numbers from 1 to 9.
- Add 9 to the southeast panel to get the numbers from 10 to 18.
- Add 18 to the southwest panel to get the numbers from 19 to 27.
- Add 27 to the northeast panel to get the numbers from 28 to 36.

Notice that the result will include the values 1 through 36, with no repeats. Now we have got the right ingredients. Also, recall that adding two magic squares results in another magic square. The only problem: we've added a square that *isn't* quite magic, so the process is not yet complete. We do obtain the same sum (111) along every row and along the two bent rows that are shaped like $\vee$ and $\wedge$. However, the column sums are either too high or too low: 84, 84, 84, 138, 138, 138. The bent rows shaped like $>$ and $<$ are likewise too high or too low by 21 (figure 8.3, right). To compensate, we'll take

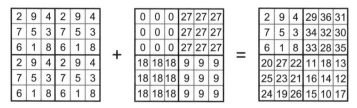

Fig. 8.3.

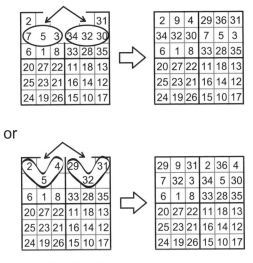

Fig. 8.4. Franklin-Strachey magic squares.

three elements from the northwest quarter and exchange them with the corresponding elements in the northeast corner. For this step, Franklin uses the middle row of the original square, Strachey the upper bent row of the original square (figure 8.4).

The row sums are undisturbed by this exchange, since reordering the elements of a row does not change its sum. Nor have those bent rows ($\vee$ and $\wedge$) been changed. All of our other troubles have been fixed. Essentially, Franklin swaps a row in the original matrix with the same row in its right-hand copy, while Strachey uses a bent row instead. The former method leads to a bent-row magic square, the latter to a straight-diagonal square, but I think the similarities are striking. Strachey's method has been celebrated by magic square enthusiasts ever since it appeared in the literature almost nine decades ago. At least some of their praise should be redirected to the humble Philadelphia electrician, who, it seems, had more than one trick in his magic bag after all.

Miraculously, Franklin's method can be made to work for order 8, 10, 12, and so on. The fact that Franklin's steps can be extended so easily to other sizes suggests that he may have examined the more general case first, which would indicate some creative use of algebra on his part. Notice that he did not have to reorder the second row; any one of the top three rows would have worked equally well. If your goal was to create a magic square of order 8 instead, you'd begin with a square of order 4; make 4 copies; add a certain matrix (described in Appendix 2); and re-order two rows in the north and two in the south. Try it and see! It can be shown using enumerative combinatorics (that is, fancy counting) that Franklin's method will generate hundreds of different magic squares of order 10, thousands of order 12 through 20, and millions for each order higher than that.

Franklin asserts that this 6-square was "more difficult to make" [than his better known squares], "though nothing near so good." It is no wonder. Not only would his 8-square method fail when applied to size 6, it can also be proved that *no other method* will work either. That is, it is impossible to fit the numbers 1 through 36 into a 6×6 grid in such a way that bent rows can be moved about with such abandon as in chapter 5. In a square this small, some of the sums are guaranteed to fail.[7] Franklin had to settle for a matrix with a little less magic, this time.

When I learned of the Franklin-Canton letter on magic, so very obscure when compared to the much earlier Franklin-Collinson letter on the same topic (chapter 5), I decided to track down the original handwritten document. The letter itself has a strange history of being ignored. Though a magic square and circle appear in early collections of Franklin's scientific correspondence, this particular letter does not. It is also absent from an edition of his collected works edited by grandson William Temple Franklin (1818), and from two more comprehensive later collections (Sparks, 1836–1840, and Bigelow, 1904). When it was finally printed (Smyth, 1905–1907), the actual squares were removed from the text! An excerpt, likewise square-free, next appeared in the definitive *Papers of Benjamin Franklin* (Vol. 4, 1961). The 6- and 4-squares finally surfaced in print with the 1968 publication of volume 12 of the *Papers*, but they went unnoticed by the mathematical community. Was there more? Where is the original document?

The *Papers* makes clear that the original manuscript was not found. A draft copy in the library of the American Philosophical

Society carried the circle but no squares. If a separate draft described the squares, it is not in evidence. In 1906, A. H. Smyth reported in his *Writings of Benjamin Franklin* that the original letter was at the "Museum of the Guild Hall, London." However, by the mid-1970s the Guildhall collections were split between the Museum of London and the Guildhall Library, and neither institution possesses the letter today. That it was sold is quite improbable; it is simply missing. According to the *Papers* (Vols. 4 and 12), the original letter was offered for sale by various galleries in 1938, 1943, and 1967, and the last time it was purchased for an anonymous buyer. (Some or all of these sales could refer to a manuscript different from the one cited by Smyth, as it was Franklin's habit to save a rough draft of each letter.)

Thus, the current whereabouts of Franklin's May 29, 1765 letter to John Canton remain a mystery. (If you are a private collector in quiet possession of this item, please step forward!) Fortunately, a facsimile of one version is preserved in the library of the Royal Society.[8] While there are no diagrams following the postscript, where they should appear, the letter is immediately preceded in the Canton Papers by not just two, but three magic squares! The third is shown in figure 8.5, retyped for clarity.

As the solid and dashed line segments suggest, this one has all of the properties of the 16-square from chapter 5, and many more besides. Some of the magic patterns that wind through this matrix are shown in figure 8.6. Every bent row and every diagonal adds to 2056; these can be shifted in the usual way, so that this square has the same 64 bent rows as in Franklin's famous 16-square, and also the 32 straight diagonals that were lacking there. (Therefore it is pandiagonal, like the 5×5 square in chapter 2.) The first or last half of any row or column adds to $\frac{1}{2}(2056) = 1028$, as in the earlier 16-square, but now we can also get 1028 by summing along the *middle* half of any row or column (that is, the middle eight numbers). If the matrix is drawn and quartered into four pieces, each is an 8×8 magic square with bent rows and straight diagonals summing to 1028. Every 2×2 block of four cells sums to $\frac{1}{4}(2056) = 512$. Notice that the entries $1, 2, \ldots, 8$ trace out a chain of knight's moves, as do many other consecutive sequences in the matrix.

16	255	2	241	14	253	4	243	12	251	6	245	10	249	8	247
1	242	15	256	3	244	13	254	5	246	11	252	7	248	9	250
240	31	226	17	238	29	228	19	236	27	230	21	234	25	232	23
225	18	239	32	227	20	237	30	229	22	235	28	231	24	233	26
223	48	209	34	221	46	211	36	219	44	213	38	217	42	215	40
210	33	224	47	212	35	222	45	214	37	220	43	216	39	218	41
63	208	49	194	61	206	51	196	59	204	53	198	57	202	55	200
50	193	64	207	52	195	62	205	54	197	60	203	56	199	58	201
80	191	66	177	78	189	68	179	76	187	70	181	74	185	72	183
65	178	79	192	67	180	77	190	69	182	75	188	71	184	73	186
176	95	162	81	174	93	164	83	172	91	166	85	170	89	168	87
161	82	175	96	163	84	173	94	165	86	171	92	167	88	169	90
159	112	145	98	157	110	147	100	155	108	149	102	153	106	151	104
146	97	160	111	148	99	158	109	150	101	156	107	152	103	154	105
127	144	113	130	125	142	115	132	123	140	117	134	121	138	119	136
114	129	128	143	116	131	126	141	118	133	124	139	120	135	122	137

Fig. 8.5. Magic square of side 16, rediscovered after two centuries.

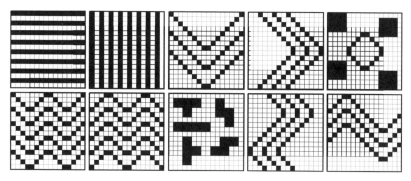

Fig. 8.6.

Thus, it *is* possible to create a 16 × 16 magic square that has just as many bent rows as in Franklin's "most magically magical square" (figure 5.8) and *simultaneously* as many straight diagonals as possible. Perhaps this example would satisfy Franklin's critics, who

(a) How do we know in advance that the magic sum in a 16×16 square composed of the first 256 positive integers must equal 2056? *Hint*: Use the 3×3 case as your guide (chapter 1). This time, however, you'll need a more sophisticated argument to avoid summing all of those numbers by hand. (The gunpowder example from chapter 3 may be useful here.)

(b) What is the magic sum in an $n \times n$ square composed of the first n^2 positive integers?

implied (however unfairly) that his bent rows were inferior to the more traditional sort of diagonal: why not include both types of pattern? But let's not be too hasty. Those critics were focused on the 8×8 case, which might be more difficult to solve. After all, there's less room to maneuver in a small square. The creation of the 8×8 magic square can be thought of as a solution to a system of 64 equations (one for each property) in 64 variables (one for each cell). Compare the 16×16 square, which involves *twice* as many equations but *four times* as many variables. It's true that some of the equations are redundant, and we are also neglecting to worry about the fact that the entries should consist of the first 64 (or 256) counting numbers. But just based on a rough argument, it looks as if the 16×16 case should be easier to conquer, since the number of unknowns is far greater than the number of conditions to be satisfied. Maybe Messrs. Hutton, Dalby, Youle, Watson, and Nulty would still remain unsatisfied, then, unless Franklin had worked out the 8×8 to their specifications.

It is not only Franklin's early followers who were discontented. Even today, authors frequently complain that Franklin's squares are not *really* magic, because the diagonals do not sum properly. Now, it's easy to flip the right-hand half of his 8-square to get straight diagonals; but that same process destroys the bent rows, so it's no real solution. One of the most widely circulated popular science magazines in the world ran a story recently that celebrated the doubly magic 8×8 squares (with both bent and straight patterns simultaneously) and challenged readers to build their own examples. Was Franklin really incapable of producing these himself? We will find out, later on in this chapter.

It is generally assumed that Franklin's magical pursuits occupied only a brief period ending in the 1750s. Contravening the conventional wisdom, we see now that Franklin was still conducting his arithmetical experiments in 1765, long after his last visits with James Logan. Five years ago, I published a research paper that considered these new squares alongside the ones we saw in earlier chapters, "completing for the first time a catalog of Benjamin Franklin's surviving magic squares." It was my hope that by presenting all of them together in one place, in a high-profile mathematics journal, I would ensure that Franklin's numerical genius would no longer be based on such a narrow amount of evidence. I was not to be disappointed. After the article was previewed at a professional meeting, it became the subject of a brief news item in the journal *Science*, and an accompanying illustration of the alternate 8-square (chapter 7) ensured that this magic square would not be ignored in the future. The 4-square appears in a new biography of Franklin written by one of his descendants. The 16-square is included in a book that accompanies the traveling exhibition *Benjamin Franklin: In Search of a Better World*. The handwritten 6- and 4-squares are reproduced in faux icing on Ben's official 300th birthday cake, a colossal layered creation over eight feet in height! And these are just the most high-profile examples. More importantly, these magic squares have spread quickly through the fraternity of amateur mathematicians around the world, a quiet but dedicated group. It seems there is no longer any danger that this particular talent of Ben's will be taken for granted any time soon.

The biggest thrill from a personal point of view came when I received a note from a very famous mathematician, who described the new 16-square as "nontrivial." This is high praise. For those unfamiliar with the mathematical usage of "trivial," I will explain by way of a timeworn joke.

A mathematics professor states a difficult theorem but neglects to justify it, saying only that the proof is "trivial." The students are collectively dumbfounded. One brave student raises her hand and then speaks.

STUDENT: Dr. Times-Table, is the proof really trivial?

The professor considers this, turns to the board, and scribbles a formula or two. He scratches his head, erases what he's written, then proceeds to cover the board with equations, gradually working himself into a frenzy. This goes on for half an hour while the students struggle to keep up. Finally, he turns to the class wild eyed, his hair frazzled, his glasses askew, the very picture of a mathematician. He then announces his conclusion.

PROFESSOR: Yep. The proof is trivial.

After 1765, the Canton-Franklin correspondence returned to the usual topic—electricity—and there are no more magic squares to be found. However, Franklin was far too busy with political affairs to spend much time involved in experimentation. In 1766, he testified before the House of Commons to argue against the Stamp Act, which was repealed within a fortnight. Soon Franklin would be appointed to represent more colonies, on his way to becoming not a Bostonian nor a Philadelphian, but an American; and moreover a citizen of the world, as he was inducted into learned societies across Europe.

It certainly seemed as if that was the end of the story.

He began corresponding with Jacques Barbeau-Dubourg, a distinguished medical doctor and translator who would prepare a French edition of Franklin's scientific correspondence. They had probably met when Franklin traveled to the Continent in 1767. Dubourg examined those famous squares of size 8 and 16, and he found them wanting. His complaint was a familiar one: they lack straight diagonals. Franklin's original reply was lost in the mail, unfortunately, but he sent a concise recapitulation to Dubourg, in which he claims to have "a square of 8, with the diagonals you required." Their letters also mention a "square of 11000" which Dubourg has created, but one in which Franklin has pointed out "some imperfection."[9] It's a curious allusion. When Franklin uses the phrase "square of 8" or "square of 16" in letters to Collinson, Canton, and Dubourg, he is plainly describing the length of one side; but when in the very same sentence he refers to a "square of 11000," surely this cannot mean that its length on each side is 11,000. While such a matrix exists, we can rule out the possibility that anyone would write out more than a hundred million entries by hand. A different interpretation is found in a detailed study of the Dubourg-Franklin relationship, where we find the claim that Dubourg constructed a magic square "with 1,100 sections."[10] This must be a typographical error. In any case, it seems that the author assumes that there are 1,100 or 11,000 *cells* contained in the matrix; but neither of these values is a perfect square, which makes the interpretation unlikely unless we assume that these two correspondents were rounding off the dimensions. There is no need for such a convoluted hypothesis, however. The truth of the matter can be found in an earlier letter of Dubourg, which makes plain that the *sum* is 11,000. All of which gets us no closer to finding the mysterious matrix.

(a) If Dubourg wrote (on average) one entry every second, and never paused to sleep or eat or do anything else, precisely how long would it take for him to fill a square matrix 11,000 cells on a side?

(b) In an earlier exercise, we found a formula that relates the size of a square to its magic sum. Can this formula be used to find the dimensions of a square whose magic sum is 11,000?

Apparently scholars have failed to locate the missing magic squares of both Franklin and Dubourg. Yet they are hiding in plain sight! In the 1773 *Oeuvres*, edited by Dubourg, we find both of them.[11] This final "lost square" of Franklin's has escaped notice by generations of historians as the result of an oversight on the part of the translator. When Dubourg translated the Franklin-Collinson letter describing that fateful visit to the home of James Logan, he

2	57	6	61	8	63	4	59
7	64	3	60	1	58	5	62
49	10	53	14	55	16	51	12
56	15	52	11	50	9	54	13
42	17	46	21	48	23	44	19
47	24	43	20	41	18	45	22
25	34	29	38	31	40	27	36
32	39	28	35	26	33	30	37

Fig. 1.

Fig. 8.7. Franklin magic square, another forgotten gem. American Philosophical Society Library.

made one important change: he replaced the classic square, which had appeared in the English *Experiments*, with the new square Dubourg had solicited from Franklin.[12] (See figure 8.7.) There was no note explaining the substitution, which is why all later editions of Franklin's works missed it. Even the most careful historian, after verifying that the text of the letter included nothing new, might reasonably assume that this 8×8 matrix must be the same one printed in the English edition of Franklin's works. It isn't. If you trace the numbers 1,2,3,... each in turn, you will see that this magic square is quite different from his others, although there are some stylistic similarities. (See chapter 9 and appendix 4.) Perhaps now the catalog of Franklin's squares is finally complete.

If we are looking for bent rows, this one has plenty—as many as his other 8-squares. If we are looking for straight diagonals, it has those too. Surely this is his greatest numerical discovery. Thus Franklin answers all critics, both past and present, who implied that he was incapable of combining both qualities in a single compact 8-square. What's more, we can use this example to help analyze Franklin's general method, in the next and final chapter. So there it is, believe it or not: another long-forgotten square from Dr. Franklin.

It turns out that Dubourg's contribution is also an 8×8 square. The sum is chosen, he says, "in honor of the 11000 virgins of our legends" [translation mine]. This would be the tale of Saint Ursula and her 11,000 handmaidens, martyred en masse in a jilted rage by Attila the Hun and his brutal compatriots. There are many variations to the legend. Maybe it wasn't Attila but some other marauder from the East. Maybe the 11,000 was a misreading of 11; it has even been suggested that this high estimate was intended to account for the suspiciously large quantity of skeletal "relics" that were unearthed in the Middle Ages. Whatever the actual details, the story was well known in medieval times. Today it is best known through its commemoration in classic song by Hildegard of Bingen, the twelfth-century nun-scientist-mystic. (Hildegard continues to be credited with miracles, and her music is still performed in concert halls. By contrast, Dubourg's magic square, every bit as consonant, has been forgotten.) What about in Franklin's time? Was Ursula's legend so well known in the English world that Dubourg could reasonably assume no further explanation would be necessary? I think so, because *Poor Robin*'s almanac for 1724 relates a joke that is predicated on just such prior knowledge:

> I have heard a story of two Papists (Zealous ones I warrant you) one a Frenchman and one an Irishman, contending about their Religion, and which of the two were the most knowing, and agreed that it should be determined in Favour of him that could name the greatest number of Popish Saints without the help of a Book, and each assuring himself of the Majority, they both agreed, that for every Saint the one named he should pull a Hair out of the other's Beard, and he that had the least Beard left, should lose the Wager. Well, to Work they went, and the Irishman proud to begin, he pluck'd a hair out of the Frenchman's Beard and Cries St. *Patrick*, the Frenchman pluck'd a hair out of the Irishman's Beard and Cries St. *Dennis*, the Irishman thinking to show his Wit, he pulls Seven Hairs out of the Frenchman's Beard, and cries the *Seven Brethren*, the Frenchman being ready prepared pulls Forty hairs out of the Irishman's Beard, and cries the *Forty Martyrs*, the

Irishman feeling it smart, pluck'd off half the Frenchman's Beard and cries the *Eleven Thousand Virgins*, the Frenchman being in a Passion, and resolving to make Reprizals, pluck'd off all the Irishman's Beard, and cried *All Saints*, at which the poor Irishman was forced to give up the Game, and yield to the other, because he had lost all that he had to play for.

A little anti-Catholic sentiment was not unusual in English almanacs. Within the same pamphlet, Poor Robin complains that there are not enough Protestant saints to fill the calendar, so he must also list some Catholic ones; but he distinguishes these by color lest we should confuse the two. Other examples abound. In *Poor Richard's* almanac for 1740, we find three prophecies credited to the ghost of Titan Leeds. (Apparently Titan's spirit entered Richard's sleeping body one night, seized control of his writing hand, and left behind several prognostications.) The first of these predicted that "J. J——n, Philomat, shall be openly reconciled to the church of Rome, and give all his goods and chattels to the chapel, being perverted by a certain country school-master." He also foretold that a certain "old friend W. B——t shall be sober 9 hours, to the astonishment of all his neighbors" and that the Bradfords "will publish another almanack in my name, in spite of truth and common sense." He must refer to rivals John Jerman and Poor William Birkett. Rather than take the high road, Jerman angrily denied the accusation, calling Franklin "one of Baal's false prophets." Franklin responded with further charges of "popery" and the "adoration of saints."[13] It was a mistake for Jerman (or anyone else) to engage Franklin in a battle of wits.

Dubourg's creation is displayed in figure 8.8, right.[14] Add any row or column and you obtain the number of Ursula's handmaidens. Diagonals work as well. Some, but not all, "bent rows" also add to 11,000. Various other Franklinesque properties are present or partially so, though it is by far inferior to Franklin's own creations. Unlike most of the magic squares we have seen, this one is not composed of consecutive integers. This is disappointing until you realize that Dubourg did begin with the values 1,2, . . . ,64 and then transformed

Fig. 8.8. Above: *The Martyrdom of Saint Ursula by Attila the Hun*, c. 1872 by Johann Schmitt (1825–1898). A mural, twenty-seven feet wide, on the arched ceiling of the former Chapel of the Ursuline Sisters of Louisville, KY. Photo courtesy of the Ursuline Sisters of Louisville. Facing page: *Quarré magique des 11000 vierges* by Jacques Barbeau-Dubourg, 1772. American Philosophical Society Library.

these numbers to get the desired sum of 11,000. How do you think he accomplished this task? (*Hint:* Find the smallest value in his square above, then locate the second and third smallest. What is the pattern?)[15]

Extra Challenge

(a) Dubourg could have used many other strategies to accomplish his purpose, transforming the values 1,2,3, . . . ,64 in an ordinary 8×8 magic square to get one with a sum of 11,000. Find one of these alternate strategies.

(b) Did he have to begin with an 8×8 magic square, or would some other size suffice?

MAGIQUES. 283

1346.	1404.	1314.	1436.	1330.	1420.	1362.	1388.
1406.	1344.	1438.	1312.	1422.	1328.	1390.	1360.
1432.	1318.	1400.	1350.	1384.	1366.	1416.	1334.
1316.	1434.	1348.	1402.	1364.	1386.	1332.	1418.
1414.	1336.	1382.	1368.	1398.	1352.	1430.	1320.
1338.	1412.	1370.	1380.	1354.	1396.	1322.	1428.
1372.	1378.	1340.	1410.	1324.	1426.	1356.	1394.
1376.	1374.	1408.	1342.	1424.	1326.	1392.	1358.

Fig. 8.8. (continued)

By the time the *Oeuvres* was published in 1773, war seemed un-avoidable, but still Franklin held out hope for reconciliation. On the one hand, he continued to pursue the colonists' interests with a steadfast (if polite) zeal. He leaked the Hutchinson letters, which brought things to a boil. (Governor Hutchinson of Massachusetts had recommended an "abridgement" of rights and "restraint of liberty." Franklin saw that these letters reached the Assembly in Boston, which touched off a firestorm of controversy.) He wrote the essay *Rules By Which A Great Empire May Be Reduced To A Small One*. Yet compared to many of his fellow colonists, Franklin was a moderate, and a latecomer to the idea of independence. Even now, he was still open to a peaceful outcome if only the motherland would consent to a fair settlement, and his willingness to negotiate was no secret. Late in 1774, he received an invitation—by way of a mutual acquaintance at the Royal Society—to visit Catherine Howe, whom Franklin later recalled as having "a good deal of mathematical knowledge." (He says little else about her, only that he "had never conceived a higher opinion of the discretion and excellent understanding of any woman on so short an acquaintance.") The stated purpose of their meeting was so that she could beat him at chess. Franklin had once been a dedicated player, though he was now "long out of practice," and he accepted the offer of a match. History does not record the victor, but they agreed to a second meeting. This time, after play, their conversation turned to mathematics, and then to the political situation. Both enjoyed themselves, and Franklin suspected nothing unusual about the meeting, no hint of subterfuge. At a later meeting, however, chess was pre-empted by a visit from Catherine's brother, Admiral Richard Lord Howe. It was Christmas Day, 1774. On behalf of himself and certain unnamed friends, the admiral appealed for Franklin's assistance, for "no man could do more towards reconciling our differences than I could if I would undertake it," as Franklin later recalled. Lord Howe expressed regret at the poor treatment Franklin had received during the Hutchinson affair, when some regarded Franklin's actions as outright treason. Howe asked Franklin to prepare an inventory of reasonable terms under which the colonies would be satisfied, without upsetting "the dignity of government." Franklin resolved to do so, and they agreed to meet again under the guise of another chess party at the home of Catherine Howe. Yet they were unable to reach agreement on terms, for the breach between England and its American colonies was too great, and war was now inevitable. In many ways, it had already begun.[16]

In the midst of threats to his personal safety, Franklin returned to Philadelphia. Sadly, his wife had died the year before, having never fully recovered from a stroke suffered four years earlier. Franklin served as a delegate to the Second Continental Congress. In the summer of 1776, Admiral Howe came to New York with an armada, joining thousands of troops that had already arrived under the command of his brother, General William Howe. Franklin, John Adams, and Edward Rutledge were sent to Staten Island to meet with the admiral. His Lordship hoped to convince the Americans to acquiesce, though he could offer no meaningful concessions. Howe was accompanied by his secretary, Henry Strachey, who by an astonishing coincidence was great-grandfather to the mathematician cited earlier in this chapter. (Henry's sons and grandsons were also gifted in mathematics. His younger son, Ralph's grandfather, published a history of Indian algebra.) When Strachey met Franklin and Adams again, six years later, circumstances would be very different; this time he led the British side in treaty negotiations following Cornwallis's surrender.[17]

Franklin served on many congressional committees. When his Committee of Secret Correspondence sent an agent, Silas Deane, on a clandestine mission to France for material assistance, Deane was instructed to approach Franklin's steadfast friend Dubourg, who (it was hoped) would provide a personal introduction to the French foreign minister. Soon, Congress sent Franklin as well, to pursue a treaty of alliance with France. He brought grandsons William Temple Franklin and Benjamin Franklin Bache along. Meanwhile William Franklin, governor of New Jersey, had been jailed for his loyalty to the Crown. The disappointed father did not use his considerable influence in order to free his son, though ironically he arranged prisoner exchanges throughout the war's duration.

Franklin's tenure as one of the American commissioners to France was spectacularly successful. This time he spent nearly nine years abroad. In general the war years displaced Franklin's philosophical speculations—he had done little in that vein since 1768, when he mapped the Gulf Stream (yet another first) and created a new phonetic alphabet. The latter invention was intended to simplify English spelling, which was even less consistent in the eighteenth century than it is today.[18] He never abandoned the idea, and as late as the 1780s he corresponded with a young Noah Webster on the topic. But while the improved alphabet was meant to make communication more transparent, as usual for the betterment of all humanity, there were times when obfuscation was the goal instead. In the

course of his wartime duties, Franklin often had to rely on encrypted messages. An eighteenth-century Freemason leader would
already have some experience with elementary ciphers; the methods used in the revolutionary war were something quite different.

One of the simplest methods of concealment, still popular as a
game among schoolchildren, is the shift cipher. Each letter of a
message may be shifted to a letter that occurs some fixed distance
later in the alphabetic sequence, so that DEATHANDTAXES becomes KLHAOHUKAHELZ if we use a seven-letter shift. It is an old
idea, in use at least as far back as the time of Julius Caesar, who
used it himself. Notice that any letter shifted past the end of the alphabet merely cycles back to the beginning (like T to A, X to E), in
the same way that a knight moving off the edge of a magic square
reappears around the other side. Since the number of possible shift
ciphers is very small, this is not a terribly secure method for important communications. Much safer is the general substitution cipher, the sort you see in newspaper cryptograms today. Here each
letter of the alphabet is replaced by another, but the scheme is not
based on a simple shift or some other easily discovered transformation. Even so, any text of sufficient length that has been produced by a substitution cipher will betray its message based on letter frequencies, unless particular care is taken to avoid common
usage. To evade this insecure feature, Franklin and his correspondents used a more sophisticated approach. One of their schemes
went as follows. First, both parties agreed on a source text to build
the key. For our purposes the preface from Franklin's 1733 almanac
will do fine. The numbers 1,2,3, etc. are used to label the letters on
the page, like this:

c	o	u	r	t	e	o	u	s	r	e	a	d	e	r	i	m	i	g	h	t
1	2	3	4	5	6	7	8	9	10	11	12	13	14	15	16	17	18	19	20	21

i	n	t	h	i	s	p	l	a	c	e	a	t	t	e	m	p	t
22	23	24	25	26	27	28	29	30	31	32	33	34	35	36	37	38	38

t	o	g	a	i	n	t	h	y	f	a	v	o	r
39	40	41	42	43	44	45	46	47	48	49	50	51	52	...

and so on through a full paragraph or page. So far we can see that 12, 33, 42, and 49 can all be used to represent the letter *a*, so it would be much harder to apply letter-frequency analysis to decipher a message like this:

22, 48, 47, 2, 8, 31, 12, 23, 10, 32, 42, 13, 24, 20,
18, 27, 47, 40, 8, 52, 6, 5, 51, 2, 1, 29, 7, 27, 6.

In our example we only bothered to label the first 52 letters of the source text, but typically the numbers ran well into the hundreds. Also, in some of Franklin's diplomatic messages only a few words or phrases were enciphered, leaving the less vital information in plaintext, thus denying the enemy additional material that would aid in decryption. Franklin's key was titled *le chiffre indéchiffrable* (the indecipherable cipher), a term often applied to similar methods.[19]

You would like to encrypt an English message composed solely of upper-case alphabetic characters, no numbers.

(a) How many different shift ciphers are possible?

(b) How many different substitution ciphers are possible? Assume that each letter corresponds to just one symbol in ciphertext. (*Hint*: How did Poor Richard calculate the number of words that can be written from a given alphabet?)

Cryptogram (substitution cipher)

MBW DVOWEM DFW EWIWF LQMBKGM RDGAM, EKF MBW NFWOWEM LQMBKGM WZSGOW. —NKKF FYSBDFC.

One of Franklin's spies was Patience Wright, an American artist working in London.[20] A contemporary and friend of Benjamin West, the greatest American painter of his era, she preferred to work in wax and was renowned for the lifelike forms she created. Among her subjects were Franklin and King George III.[21] On May 4, 1779,

Franklin drafted a letter to Mrs. Wright. It does not concern espi-onage, and is not encrypted. In the postscript, he relates a playful idea from his grandson and private secretary, who suggests that her wax figures may be transported to America more safely and economically if they travel as regular passengers instead of being boxed up as luggage. Only a joke, William's suggestion plays off her well-deserved reputation for flawlessly capturing the images of the famous and powerful. He concludes that "all the world will wonder at your clemency of Lord N[orth], that having it in your power to hang or send him to the lighters, you had generously reprieved him for transportation." A draft copy of this letter rests today at

Lon: Mag.

M.ᴿˢ WRIGHT.

Published as the Act directs Dec.ʳ 1. 1775.

Fig. 8.9. Wax modeler at work: *Patience Lovell Wright (With the Head of Benjamin Franklin)*, unidentified artist, engraving, 1775; published in *The London Magazine*, Dec. 1, 1775. *National Portrait Gallery, Smithsonian Institution.*

the Library of Congress. On the reverse is written "Postscript to Mrs. Wright, Miscellaneous Papers," and below that, in faded pencil, we find a small grid partly filled with numbers. Since the microfilmed version available in libraries does not legibly reproduce the values in that matrix, I was granted special permission to view the original. In person, with a magnifying glass, you can barely make out twelve values in the upper three rows of a 4×4 matrix, with the last row blank, possibly crossed out (figure 8.10).[22]

It appears to be a brief and failed experiment. For all we know, each of Franklin's great magic squares was preceded by a hundred such "rough notes," none of which were worth keeping. Figure 8.10 shows how the numbers from 1 to 12 are placed in a long chain of knight's moves.[23] It also shows how diagonal neighbors are paired in sums of 12 (for the first two rows) or 14 (for the second and third rows); these diagonal pairings recall the third of our five templates for building 4×4 magic squares (figure 3.6).[24] Consequently all

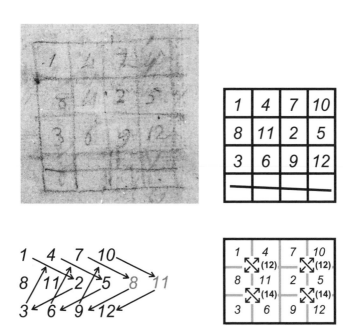

Fig. 8.10. Upper left: Sketch from Franklin Papers in the Library of Congress. Retyped for clarity, upper right.

three 2×2 blocks that occur in the top two rows have sum 24, and the three blocks in the second and third rows have sum 28. (Most of Franklin's squares are arranged to have a constant block sum.) If you connect the pairs {1,12}, {2,11}, and so on each by a line segment, you obtain an asterisk similar to the ones we drew over Franklin's 6-square. Finally, it appears that the author of this little matrix may have simply listed 1,2,3 down the first row, 4,5,6 down the next, and so on, then swapped the two halves of row 2—the same way Franklin did in the 6×6 case. (Another way to look at it is that the arithmetic sequences 1,4,7,10 and 2,5,8,11 and 3,6,9,12 have been placed in these three rows, but the second row starts in the middle and cycles around.) Thus, there are at least vague similarities to the squares we have seen before. We have not found another magic square here, but this is something more than a random set of numbers. It appears to be the only evidence of Franklin's rough notes experimenting with magic squares in an uncorrected state. This extends further the time period in which Franklin paid at least occasional attention to magic squares, to four decades or more. At the advanced age of 73, on the verge of his final decade, those little recreations still were not forgotten.[25]

Franklin was then living in Passy, at just a slight remove from the Parisian culture that so adored him.[26] Even there, far from home, he ordered printing supplies and set up a working press. It is often said that no matter how many roles Franklin played, offices he held, and fields of knowledge he contributed to, he was always at heart a *printer*.

The printing trade has many mathematical aspects. In Franklin's time, every line of a written work had to be arranged one letter at a time on a composing stick before going to a galley. The letters and spaces of a single row on the stick are lined up like the entries of a vector (a row or column of a matrix). Successive lines of text fail to conform to the matrix analogy, however, since the varying width of different letters ensures that these are not likely to fall in columns on the printed page or the composing stick. (Franklin learned its use as an apprentice, and he was employed in the composing room of two different printers during his first trip overseas in the 1720s.)

Some arithmetic is required to determine the number of lines that will fit on a page, and that value is partly determined by font sizes. Those were distinguished by a confusing set of names (pica, double pica, primer, etc.) instead of the numerical point size used today.

In font design, a single letter was often sketched against a rectangular grid so that it could be easily reproduced. One author notes that "under the reign of Louis XIV, as the poor and the mad were confined, so letters were locked into the prison of the grid," a square background made up of 64 small squares (like a chessboard, or one of Franklin's magic squares).[27] The letter E might be reproduced by following a diagram like the one given in figure 8.11. Joseph Moxon's *Rules of the Three Orders of Print Letters* (1676), on the other hand, explains that each letter is "compounded of geometric figures" that can be created using the basic tools of ruler and compass. Moxon was a printer who wrote and published books on astronomy, geometry, architecture, and many other topics; his mathematical dictionary was a popular reference work. A very rare copy of his *Mechanick Exercises*, a printer's reference, was in Franklin's personal library.[28] Moxon's process for correctly drawing the capital letter C includes these instructions: "Set your Compasses to 15, and placing one Foot in Parallel 27 Erect 15 . . . with the other describe a Circle: Cut off half a Stem . . . of this Circle on the right hand with a Perpendicular line. . . ."[29] (The complete instruction is many times longer than this excerpt.) Whenever not in

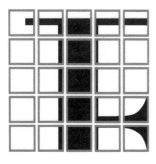 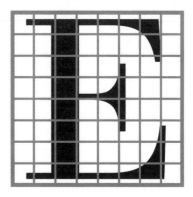

Fig. 8.11.

use, type is kept in a rectangular storage grid called a "case" (not to be confused with the *cases*, or individual cells, of a *carré magique*); standard typecases with 7×7 submatrices appear as illustrations in Moxon's *Exercises* and Diderot's *Encyclopédie*. But all this shows is that the matrix, like mathematics itself, is everywhere. More interesting is the process by which a single sheet is folded and cut into many pages, which we examine next.

Consider the technical terms used to describe the size of a book. Almanacs like the original *Poor Richard* and the *Ladies Diary* are classified as octavo or sextodecimo (sometimes abbreviated 8vo, 16mo). *Poor Richard Improved* was in duodecimo. Recall that Stifel's *Arithmetica Integra* (chapter 5) was in quarto. These designations indicate how many times a large sheet has been folded before being sewn and cut into individual pages. Naturally, one must be able to predict in advance where each folded sheet will land in the final product, lest a page appear out of sequence or upside down. And so a printer uses a master pattern to guide that process. One such plan, taken from a printer's handbook, is shown in figure 8.12.[30]

Imagine that the upper half of this diagram is sketched on one side of a sheet, and the lower half of the diagram sketched on the back. This is done in such a way that 1 and 2 appear in the same location on opposite sides of the sheet, as do the other front-back pairs that appear in any book (such as the one you are now reading). Folding the sheet in half along a horizontal fold will ensure that page 1/2 lies atop page 3/4. Next fold along the vertical, so that pages 1–4 are followed immediately by pages 5–8. After three more folds, the pages are all in order, ready to be sewn and cut. Altogether we have bisected the original sheet five times, doubling the number of layers each time until we get $2 \times 2 \times 2 \times 2 \times 2 = 32$ *leaves*. A book produced in this fashion is classified as 32mo. (An octavo sheet would be folded three times, so that a single sheet becomes $2^3 = 8$ smaller sheets.)

What I find interesting is that we see in this diagram the numbers from 1 to 64, arranged in an 8×8 matrix so that adding across any row produces the same sum of 260. That's because complementary

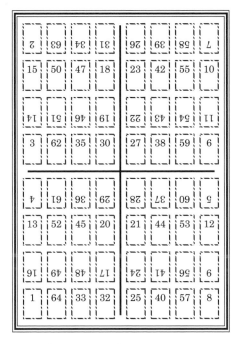

Fig. 8.12. Folding scheme.

pairs that add up to 65, like $1 + 64$ or $37 + 28$, are always next to each other within the same row. For an entirely different reason, all eight of the $\vee$-shaped bent rows add up to 260, just as in Franklin's three 8×8 magic squares. Many of the 2×2 blocks add up to $\frac{1}{2}$ (260), half of them in fact, and the half-rows also add up to 130. This innocuous printer's device bears many similarities to Franklin's squares, right down to the convenient placement of complementary pairs. His careful eye could scarcely fail to notice the mathematical regularity of the matrix above, and his inner bookkeeper would surely enjoy the balance within. As for usefulness, calculating a bent row sum provides a double-check that the diagram has been reproduced properly, akin to a "check digit" in the Universal Product Code that nourishes your local supermarket scanner. (*How many other ways, in addition to the standard one described above, can you fold a 4×8 sheet down to the size of a single page, if only horizontal and vertical halving is allowed?*)

Paper folding is a highly mathematical topic, one that is just as rich as the straightedge-and-compass constructions from your high school geometry class. I don't know of any specific study of the printer's templates, however. Cells that begin far apart may become adjacent pages, while cells that are adjacent on the original sheet may end up many pages distant from one another when the folding is done. In this way it resembles a finite analog of the "horseshoe maps" that appear in the study of chaos theory.

Speaking of paper folding, here is a neat little fact. If you fold Franklin's most famous 8-square twice, a funny thing happens: it becomes a magic cube! Write the matrix on a sheet of paper, and be sure to use a thick enough pen so that the numbers show through on the back as well. Fold down, so that the top half rests on top of the bottom half. (If you think of it as a world map, the first fold corresponds to the equator.) Next fold the left half over on top of the right side. If you cut along the folds without disturbing the position of the four quarter sheets, you get the cross sections of a magic cube. Each line of four cells along the length, width, or depth of the cube sums to 130. The top and bottom faces are "Franklin magic squares": connect any two adjacent corners by a bent row, and it will sum to 130. On the front and top faces, and on every slice parallel to these faces, there are at least two bent rows that can be shifted about in the usual way, all of them summing to 130. As an exercise, you can trace these sums back to their placement in the original square to see where the magic came from. A Franklin magic cube of order 6 was contrived a century ago by H. M. Kingery,[31] and I have just received word from Arsène Durupt of Auxonne, France, attesting to the existence of $8 \times 8 \times 8$ magic cubes that carry many of the Franklin properties. Just how many bent rows can be packed into an $n \times n \times n$ cube remains a mystery.

Though his printing press brought some of the familiarity of home, by 1785 it was time to depart. Congress finally approved the retirement of this aging, homesick statesman who had several times begged to leave on account of his ill health. On July 12, he and his grandsons left Passy for the coast, on what would be a week's journey. Only days from setting sail, they received a delegation from the academy of Rouen. At this last ceremony, a director of the academy presented Franklin with a magic square encoding his name. It was an appropriate honor for the man who made numbers do his bidding.[32]

Notes

1. Letter to Cadwallader Colden, Oct. 11, 1750. *Papers*, 1961, Vol. 4, pp. 67–68.

2. The measurements are given in Franklin's letter to Giambatista Beccaria, printed in *Experiments and Observations*. A work order with additional data is found in *Papers*, 1966, Vol. 10, pp. 180–182. The armonica was also reputed to induce premature labor. Luckily, we learned in chapter 2 that the 3×3 Saturn square provides a ready remedy for any complications.

3. The general process is described in many places, for example Siegmund Günther, *Vermischte Untersuchungen zur Geschichte der mathematischen Wissenschaften*, Leipzig: Teubner, 1876. An explicit construction of the square of order 24 may be found in Charles J. Jacobs, "A Reexamination of the Franklin Square," *Mathematics Teacher*, Vol. 64, 1971, pp. 55–62. The main idea is described at http://www.pasles.com/Franklin.html.

4. Letter of Benjamin Franklin to Achille-Guillaume Lebègue de Presle (incidentally, the personal physician to Jean-Jacques Rousseau), Oct. 4, 1777, *Papers*, Vol. 25, 1986, pp. 25–26 (quote appears on p. 26).

5. 1918 letter from Ralph Strachey to W. W. Rouse Ball. The attribution to Strachey is repeated in subsequent editions of the *Recreations*, including the most recent: W. W. Rouse Ball and H.S.M. Coxeter, *Mathematical Recreations and Essays*, 13th edition, Mineola, N.Y.: Dover Publications, 1987.

6. Charles Richard Sanders, *The Strachey Family, 1588–1932: Their Writings and Literary Associations*, New York: Greenwood Press, 1968.

7. Pasles, 2001 and 2006.

8. *Canton Papers of the Royal Society* (MS 597–599), letter Ca. 2.21.

9. Letter of Franklin to Jacques Barbeau-Dubourg, Dec. 26, 1772. Smyth, *Writings*, Vol. 5, pp. 463–465.

10. Alfred Owen Aldridge, "Jacques Barbeau-Dubourg, a French Disciple of Benjamin Franklin, *Proceedings of the American Philosophical Society*, Vol. 95, No. 4, 1951, pp. 331–392.

11. Jacques Barbeu-Dubourg, *Oeuvres de M. Franklin*, Paris, Vol. II, 1773.

12. The particular image reproduced here is actually from the German printing of Dubourg's collection: *Des Herrn D. Benjamin Franklin's ... Sämmtliche Werke: Aus dem Englischen und Französischen übersetzt*, Dresden, 1780, Zweiter Band [Vol. 2]. The magic square is Taf. 3 Fig. 1. The same magic square is in the French version, of course, but the numeric font used in that image is smaller and slightly more difficult to reproduce.

13. *Poor Richard* 1740, 1742, and *The American Almanack* for 1741. As both titles were published by Franklin, it may seem that the two authors were acting in concert; but Jerman angrily took his business elsewhere in direct response to his squabble with "*R——S——rs* alias *B——F——n*", according to *The American Almanac* for 1743.

14. *Oeuvres*, Vol. II, p. 283.

15. The values are 1312, 1314, 1316, and so on up to 1438. Dubourg doubled every entry and then added 1310.

16. Benjamin Franklin to William Franklin, March 22, 1775, *Papers*, 1978, Vol. 21, p. 540.

17. Sanders, 1953, especially pp. 184, 248. Some details of the meetings are found in David McCullough, *John Adams*, New York: Simon & Schuster, 2001; and James Srodes, *Franklin: The Essential Founding Father*, Washington, D.C: Regnery, 2002.

18. Benjamin Vaughan, ed., *Political, Miscellaneous, and Philosophical Pieces*, London: Printed for J. Johnson . . . , 1779. The alphabet was actually used in correspondence between Franklin and Mary Stevenson Hewson (*Papers*, 1972, Vol. 15, pp. 173, 215–216).

19. The cipher used by Franklin and C.G.F. Dumas is described in Fred B. Wrixon, *Codes, Ciphers & Other Cryptic & Clandestine Communication*, Black Dog & Leventhal Publishers, 1998. Examples of this type can be found in correspondence at the American Philosophical Society Library and the Historical Society of Pennsylvania. See Edmund C. Burnett, "Ciphers of the Revolutionary Period," *The American Historical Review*, Vol. 22, No. 2, 1917, pp. 329–334.

20. Charles Coleman Sellers, Patience Wright, *American Artist and Spy in George III's London*, Middletown, Conn.: Wesleyan University Press, 1976.

21. She also modeled Reverend George Whitefield (chapter 4) and Cadwallader Colden (chapters 1 and 5).

22. Letter of Franklin to Patience Wright, May 4, 1779, *Benjamin Franklin Papers in the Manuscript Division of the Library of Congress: The Franklin Papers, Miscellaneous, Vol. III, 20 October 1778–1 May 1781*, No. 525 (verso).

23. The figure shows just one way of tracing these knight's moves. This can also be accomplished without jumping off the edge of the square.

24. Every single one of Franklin's magic squares involves the clever placement of complement pairs. Instead of k and $16-k$, here he uses k and $12-k$ or k and $14-k$.

25. In my zeal to learn whether other late-era magic squares were lurking in the letters, I examined over 100 feet of microfilm on either side of May 4, 1779, but sadly I must report that these documents are "square-free."

26. Today, Passy is part of Paris.

27. Georges Jean, *Writing: The Story of Alphabets and Scripts*, New York: Harry N. Abrams, 1992, p. 102.

28. Edwin Wolf 2nd, ed., *Legacies of Genius: A Celebration of Philadelphia Libraries*, Philadelphia: Philadelphia Area Consortium of Special Collections Libraries, 1988.

29. Joseph Moxon, *Regulae Trium Ordinum Literarum Typographicarum*, London: Printed for Joseph Moxon, 1676.

30. Thomas F. Adams, *Typographia, or, The printer's instructor: a brief sketch of the origin, rise, and progress of the typographic art, with practical directions for conducting every department in an office, hints to authors, publishers, & c.*, Philadelphia: L. Johnson & Co., 1854.

31. Andrews, 1917.

32. Jared Sparks, *The Life of Benjamin Franklin, Revised Edition*, Philadelphia: Childs & Peterson, 1859, Appendix VI. I have examined the corresponding entries from the diaries of both grandsons and find no further information on the fate of that award. Franklin recalls it as "a magical square, which I think he said expressed my name. I have perused it since, but do not comprehend it." (Most likely this was an alphanumeric representation, like those used on talismans centuries earlier.) A mountain of medals and citations were awarded to Franklin in his lifetime, and like many of them, this one has gone missing.

9 Legacy

Will somebody please explain to me what these magic
numbers are?
    —Riley Poole to Benjamin Franklin Gates, in *National
       Treasure* (2005)

A renegade scholar, expert in the analysis of historical
documents, claims to have found a secret message planted in the
Declaration of Independence. The perpetrator was Benjamin
Franklin himself. Is this just a movie, or is it real life?

In the motion picture *National Treasure*, Nicholas Cage finds just
such a clue. One thing leads to another, and with further assistance from the Silence Dogood letters, he embarks on a quest for
hidden Masonic riches. (Sure, why not?) By coincidence, in 1984—
in real life, not Hollywood—a lone researcher claimed to have discovered a magic square encoded in one copy of the Declaration.
Rather than leading to the treasure of the ages, this find was intended to support his argument that Franklin, not Jefferson, was
its primary author.[1] Historians were thoroughly unconvinced, and
I must say that the purported mathematical evidence is equally
tenuous. What this story illustrates, however, is that the Franklin
magic squares continue to amaze, inspire, and tantalize, even driving an amateur historian's fable about the very foundations of our
nation.

When Franklin returned to America in 1785, the time for mathematical recreations had ended. As far as we know, his thoughts never again turned to magic squares. His final years were quiet, in the main, but not entirely uneventful. His key role at the Constitutional Convention is well known. He also served as President of Pennsylvania, and then as President of the Pennsylvania Society for Promoting the Abolition of Slavery, though his appeals to both houses of Congress on the issue went unrewarded. His donation of two hundred pounds led to the founding of Franklin College, ancestor of Franklin & Marshall College. He invented a device for extending one's reach, for the purpose of retrieving and returning books stored on high shelves. But for the most part his time was spent with family and friends at home, a deserving recompense for the first and greatest American. There he died, just two and a half blocks from the State House, where he had practiced his magic in mathematics and politics. His funeral in 1790 was attended by twenty thousand, in a city whose entire population numbered only twenty-eight thousand. Counting on the power of compound interest, he had left bequests to Philadelphia and Boston to support loans for young tradespeople; the relevant codicil to his will calculates the gifts' future value based on a reasonable 5% annual rate. It was left for others to continue Franklin's magical legacy, which would be so eclipsed in the public mind by his celebrated accomplishments in other arenas.

(a) Franklin's bequest of £1,000 to Philadelphia was to fund loans at 5% compounded annually. How much did he expect the fund to be worth after one hundred years?

(b) After 100 years, £100,000 was to be diverted for other purposes. If the remaining amount continued to increase at the same rate as before, what would be the value of the fund after another hundred years?

Professional academics and amateur mathematicians alike have contributed to this effort. The late D. N. Lehmer, one of the great twentieth-century number theorists, expanded Franklin's method to other sizes and investigated general questions regarding construction and enumeration.[2] Maya Mohsin Ahmed, a recent Ph.D. from the University of California–Davis, has done the most extensive work in this regard, finding a convenient basis—an economical set of building blocks—from which Franklin squares of orders 8 and 16 may be created.[3] Jacques Bouteloup, a textbook author from Rouen (where Franklin was honored by a gift of a personalized magic square upon his departure from France in 1785) has generated examples that are inspired by Franklin's own.[4] Other contributors are physicists, appropriately enough, such as Ray Hagstrom (Argonne National Laboratory) and Peter Loly (University of Manitoba), both of whom have worked to extend Ahmed's research.[5] In 2004, Loly and two students, Daniel Schindel and Matthew Rempel, calculated the number of 8×8 squares that are possible under a reasonable definition of "Franklin magic."[6] (There are more than a million of these, in case you were wondering.) Hagstrom has studied matrices that combine the bent-row concept with other traditional designs, counting the number of squares whose side length is a multiple of four.[7] The present author has investigated both odd and even orders. From those humble beginnings at George Brownell's school for writing and arithmetic, and his self-study of Cocker's mathematical primer, Franklin instigated a loosely organized modern research program involving linear algebra, group theory, combinatorics and number theory, in the same way that his simple but inspired discoveries in electricity are partly responsible for the modern age (if on a vastly different scale).

Another researcher is Gil Lamb, an electrical engineer who piloted Beaufort and Wellington bombers with the RAF Coastal Command in World War II. Now living in Thailand, he has written a book-length treatise on the construction of magic squares, including a whole chapter on Franklin squares. Then there is Edward W. Shineman, Jr., a retired pharmaceutical executive (with an accounting background, incidentally) whose magic oeuvre includes rectangles, diamonds, alphabetic monograms, and squares à la Franklin. Don Morris, a retired tool and die maker, has shown that the anonymous square from chapter 6 can be constructed using a method similar to most of Franklin's own squares.[8] I think that Ben Franklin, who had to learn mathematics (and so much else) on his own, would be pleased to see that his magical legacy has been continued by so many who are not professional mathematicians.

Is there something about this field that confers longevity on its practitioners? Both Lamb and Shineman are in their nineties. The late James Warbasse, Jr., a direct descendant of Franklin's sister, was a medical doctor by vocation, but he experimented with his indirect ancestor's squares until well into his eighties.[9] Franklin himself made it to eighty-four, more than twice the life expectancy for that time and place.

If to my Age there added be
One Half, one Third, and three times three;
Six Score and ten the Sum you'd see,
Pray find out what my Age may be?
 —"Arithmetical Ænigma" in the *Ladies Diary* for 1707

One of the most accomplished "followers of Franklin" was Clarence C. Marder, an amateur mathematician who authored an extensive study of the Franklin squares in 1940. A rarity in American libraries and virtually unknown elsewhere, this is a collection of dozens of matrices patterned after the Franklin style, including one that commemorates the Declaration of Independence (figure 9.1). Rows, columns, and bent rows sum to $\frac{1}{4}(1776)$. Like Dubourg, Marder uses nonconsecutive entries to obtain his arithmetic goal.

But just who was Clarence Marder? Biographic information is difficult to come by. It turns out that he worked in the printing trade, the second generation of his family to be so employed, which may explain the fascination with Franklin. If composing type for the printed page helped launch Franklin on his magical journey, it did no less for Marder. He is co-credited with designing the still-popular COPPERPLATE GOTHIC font around 1905; his colleague was F. W. Goudy, one of the most prolific typographers of all

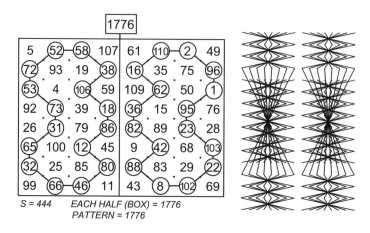

S = 444   EACH HALF (BOX) = 1776
PATTERN = 1776

Fig. 9.1. Top left: Marder's Franklin-style "1776" magic square; top right: Bragdon's "magic lines" ornament tracing the path of entries in the Franklin 16-square; bottom: Bragdon's elevator design based on the magic lines of a different square. (Some of Agrippa's occult symbols were derived in the same fashion.)

time. In the early decades of the twentieth century, Marder managed a type foundry, and between 1940 and 1941 he published *The Intrinsic Harmony of Number*, a series of monographs which included his study of Franklin.[10] That title was borrowed without attribution from philosopher-architect and cult figure Claude Bragdon, whose 1928 essay "Man: The Magic Square" asserts that magic squares "are conspicuous instances of the intrinsic harmony of number, and as such serve as an interpreter to man of that cosmic order which permeates all existence." (Bragdon suggests that they may even indicate "the operation of some supernal intelligence.")[11]

Marder mimicked Franklin's squares by combining two simpler matrices into a single magic square. He was not unique in that regard. The many amateur and professional mathematicians who examined Franklin's magic squares have remained largely unaware of one another's existence, and Marder was no exception. The reason for this disconnect is that, on the few occasions they publish, their work usually finds a place in magazines and books that are not a part of the mathematical mainstream; for example, Marder's work was printed by a tiny press, then quickly forgotten. Consequently, many of their ideas have been rediscovered independently over and over again. Those who come closest to uncovering Franklin's own steps have, like Marder, proceeded according to the recipe we will describe in this chapter. It is a method that was decoded by one of the earliest "followers," whom we met back in chapters 6 and 7. But first, it will be useful for us to step back into the present.

As I write this, in the opening years of the third millennium, one of America's favorite leisure pastimes is the sudoku puzzle (figure 9.2). The goal is to fill the empty cells with numbers so that each of the nine rows, nine columns, and nine boldfaced 3×3 blocks includes the values 1 through 9. (Try it!) Once a sudoku puzzle is solved, the sum of any row or column is 45, which means that this addictive puzzle is a variety of magic square. To be more specific, it is a *Latin square*, an $n \times n$ matrix whose entries come from a set of n symbols, each of which appears precisely one time in each row and column.

5		7		6			8	
	4		3			6		1
		2		9	4		7	
	9		2		5			6
		1	6				9	7
4	5			7		8		
3				1	8	2		
		4			6	9		8
8	6		4				1	

Fig. 9.2. Sudoku.

The sudoku craze is so pervasive in the United States—the puzzle appears as a regular feature in hundreds of daily newspapers—that it has been the subject of dozens of news stories in the print and television media. The publisher of the Japanese magazine generally credited with its invention has acknowledged that it is based on a "number place" puzzle that appeared in an American magazine in 1979.[12] Be that as it may, the more general concept of a Latin square goes back much further. The most frequent attribution is to Franklin's Swiss contemporary Leonhard Euler, perhaps the greatest mathematician of all time. According to the *New York Times*, ". . . Euler invented Latin squares, the basis of sudoku, in 1783," and similar claims are made elsewhere.[13] The truth is subtler. While he laid the groundwork for a mathematical theory of these matrices, the notion is not exclusively Euler's. Eighty years earlier, François-Guillaume Poignard (in what is now Belgium) hit on the idea of placing n copies each of the numbers 1 through n (or 0 through $n-1$) in an $n \times n$ matrix so that no single row, column, or diagonal allows for any repeats.[14] This was simply a Latin square under another name, with a diagonal condition attached.

In 1723, a new edition of Ozanam's *Recreations* took the idea further with a novel challenge. To complete this exercise you will need an ordinary deck of playing cards. Set aside the cards numbered 2 through 10, which won't be needed. Now try to place the remaining "court cards" and aces in a 4 × 4 array so that each rank and each suit appears precisely once in each of the four rows, four columns, and two diagonals. Read no further until you have tried it!

One such arrangement is shown in figure 9.3. Your own answer may be different. In either case, if you consider only the ranks (or the suits) alone you get a Latin square, indeed a "diagonal Latin square." These separate components are said to be *orthogonal* Latin squares, because when superimposed, every ordered pair appears precisely once. For example, "ace" is only paired once with "heart" in the combined matrix.

The combined matrix formed in this way is sometimes called a *Greco-Latin square*, after Euler's practice of using Latin letters for the first component and Greek letters for the second. In 1782/3, Euler described the *36 Officers Problem*: Members of six regiments, representing six ranks within each regiment, are to march in square formation. No regiment or rank may be represented twice in any row or column. If there were four regiments and four ranks, we know that this can be done. But what Euler conjectured was that the 6 × 6 version is impossible, that no Greco-Latin square of order 6 exists. That he was correct would not be proved until 1900. (He also believed that no orthogonal pair could be found in orders 10, 14, 18, and so on, but this was the rare instance in which he erred;

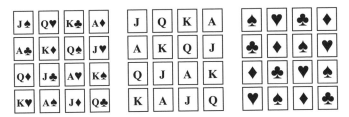

Fig. 9.3. Left: One possible answer to the 16-card square challenge. Center and right: The same answer separated by rank and suit components. Each is a diagonal Latin square.

the *only* orders in which orthogonal Latin squares cannot be found are 2 and 6. This was finally proved in 1960!)[15]

> Prove that no Greco-Latin square of order 2 exists, then find a Greco-Latin square of order 3.

The playing-card array provides a recipe for constructing a more traditional magic square. Imagine replacing J,Q,K,A by 0,1,2,3 and ♠,♥,♣,♦ by 0,1,2,3 as well. Each component matrix is now a numerical Latin square. When combined, they produce a magic square (figure 9.4). The figure at left shows the two-digit pairs, which we may interpret naively as nonconsecutive decimal values. With a little more sophistication, these may be viewed instead in base four. Whereas "base ten" digits may be read as ones, tens, hundreds, etc., "base four" digits denote ones, fours, sixteens, sixty-fours, and so "forth." In this way, the pair 3 2 (for example) represents 3 fours plus 2 ones, or fourteen. Under this interpretation, the playing-card array becomes figure 9.4, center. Finally, we can add 1 to all entries so that they run from 1 to 16 instead of 0 to 15.

Shortly after Poignard's publication, the French polymath Philippe de la Hire noted that the same strategy can be made to work even if the component matrices are not quite Latin squares. Euler would make the same observation when he first broached the subject in 1776. He used the following illustration. The first matrix on the next page includes four *a*'s, *b*'s, *c*'s, and *d*'s, though it is not a Latin square.

0 0	1 1	2 2	3 3
3 2	2 3	1 0	0 1
1 3	0 2	3 1	2 0
2 1	3 0	0 3	1 2

0	5	10	15
14	11	4	1
7	2	13	8
9	12	3	6

1	6	11	16
15	12	5	2
8	3	14	9
10	13	4	7

Fig. 9.4. Left: Base 4 numerals. Center: Converted to decimal. Right: Each value increased by 1.

As long as our numerical assignments for these letters satisfy $a + d = b + c$, we still get a magic square. If you transpose the matrix (that is, flip it over, keeping upper left and lower right corners inert) and replace Latin letters with Greek, you get the second matrix below. It's orthogonal to the first one, fortunately, so we'll get no repeated entries in our final result. The Greco-Latin square is shown next. Euler equates a,b,c,d with 0,4,8,12 and $\alpha,\beta,\gamma,\delta$ with 1,2,3,4 to get the final answer below, right. (This process is equivalent to the method we applied to the card-square a moment ago.) The same idea works for all even orders, and a different trick accounts for the odds. Assisted by this insight, he was able to show that magic squares can be constructed in every order bigger than 2.[16]

a	a	d	d
d	d	a	a
b	b	c	c
c	c	b	b

α	δ	β	γ
α	δ	β	γ
δ	α	γ	β
δ	α	γ	β

$a\alpha$	$a\delta$	$d\beta$	$d\gamma$
$d\alpha$	$d\delta$	$a\beta$	$a\gamma$
$b\delta$	$b\alpha$	$c\gamma$	$c\beta$
$c\delta$	$c\alpha$	$b\gamma$	$b\beta$

1	4	14	15
13	16	2	3
8	5	11	10
12	9	7	6

Given only the result, you can always reverse these steps to recover the numerical component squares if they exist. (Most magic squares cannot be produced by this process, however.) First subtract 1 from each entry, then divide by 4 (or by 5 for a 5-square, and so on). List all of the quotients in one matrix and all of the remainders in another, and you will have found both component squares. Nulty, Marder, Lamb, Morris, and others have all explained the Franklin squares based on this same process. *Apply this technique to the top two rows of Franklin's famous 8-square (chapter 5). Based on these two rows alone, what kind of component squares do you expect?*[17]

The first person to discover that Franklin built his matrices in this manner—at least the first one to leave a written record—was Joseph Youle, a mathematics teacher and schoolmaster who headed the Boys' Charity School in Sheffield in the early part of the nineteenth century (see chapter 6). In 2005, I was asked to organize a Franklin-themed course for a mathematics conference in Albuquerque. As fate would have it, the library at the University of New Mexico holds what is probably the only copy in America of Youle's treatise on magic squares. Thus my journey from Pennsylvania to the Southwest was doubly exciting. There I was able to learn that one of the very first "followers of Franklin" also adhered to the method of component squares.

We now have three different 8-squares to consider, more evidence than was ever available before. (See chapters 5, 7, and 8.) This will provide further confirmation that Franklin applied the same principle just described, well before Euler and—as far as we can tell—independently of la Hire. Below you will find the quotients and remainders for these three magic squares.

FRANKLIN VISITS JAMES LOGAN, 1750 (Chapter 5)

Quotients

6	7	0	1	2	3	4	5
1	0	7	6	5	4	3	2
6	7	0	1	2	3	4	5
1	0	7	6	5	4	3	2
6	7	0	1	2	3	4	5
1	0	7	6	5	4	3	2
6	7	0	1	2	3	4	5
1	0	7	6	5	4	3	2

Remainders

3	4	3	4	3	4	3	4
5	2	5	2	5	2	5	2
4	3	4	3	4	3	4	3
2	5	2	5	2	5	2	5
6	1	6	1	6	1	6	1
0	7	0	7	0	7	0	7
1	6	1	6	1	6	1	6
7	0	7	0	7	0	7	0

Look across each row of the quotient. At first glance, it appears that this may be a Latin square. The columns, however, follow a muse of their own. The reverse is true for the remainder square. Each component square alone is a magic square in the style of Franklin, except that entries repeat. With smaller numbers involved, it is much easier to check that rows, half-rows, bent rows, blocks, and so on all have the correct sum without the assistance of a calculator; and too it must have been easier for their creator to construct them, if this is the approach he used. The slightly trickier part is ensuring that the matrices so constructed are orthogonal, that when combined they produce precisely the values 1,2,3, ... 64 with no repeats. Franklin seems to have had no difficulty in doing so. Since each component alone possesses the desired properties, so does their "linear combination." (The converse is not true, in general.) Notice that the columns of the quotient square are the same as the rows of the remainder square, just in a different order.

FRANKLIN PAPERS, UNDATED (Chapter 7)

Quotients

2	5	3	4	2	5	3	4
3	4	2	5	3	4	2	5
4	3	5	2	4	3	5	2
5	2	4	3	5	2	4	3
6	1	7	0	6	1	7	0
7	0	6	1	7	0	6	1
0	7	1	6	0	7	1	6
1	6	0	7	1	6	0	7

Remainders

0	6	5	3	4	2	1	7
7	1	2	4	3	5	6	0
0	6	5	3	4	2	1	7
7	1	2	4	3	5	6	0
0	6	5	3	4	2	1	7
7	1	2	4	3	5	6	0
0	6	5	3	4	2	1	7
7	1	2	4	3	5	6	0

The remainder square, above right, looks quite like the ones we have already seen. The quotient square follows a different pattern: while each column contains the entire entry set, each row consists of a repeating pattern of four entries. I will refrain from comment on the third example, below, except to notice its affinity with the first two.

"LOST" SQUARE TO DUBOURG, 1772 (Chapter 8)

Quotients

0	7	0	7	0	7	0	7
0	7	0	7	0	7	0	7
6	1	6	1	6	1	6	1
6	1	6	1	6	1	6	1
5	2	5	2	5	2	5	2
5	2	5	2	5	2	5	2
3	4	3	4	3	4	3	4
3	4	3	4	3	4	3	4

Remainders

1	0	5	4	7	6	3	2
6	7	2	3	0	1	4	5
0	1	4	5	6	7	2	3
7	6	3	2	1	0	5	4
1	0	5	4	7	6	3	2
6	7	2	3	0	1	4	5
0	1	4	5	6	7	2	3
7	6	3	2	1	0	5	4

These examples bear a striking resemblance to an 8-square from Philippe de la Hire (see appendix 3). The la Hire/Franklin/Euler method can be summarized as follows for order 8: (1) Using eight zeros, eight ones, eight twos, and so on up to sevens, fill an 8×8 matrix so that each row and each column sums to 28. If you desire additional properties (like bent rows and 2×2 half-sums), incorporate these into your matrix too. (2) Repeat step 1 to create a second matrix, but you must do so in such a way that the second matrix is orthogonal to the first. This is not as easy as it sounds!

The 16-squares follow the same procedure, though Franklin's 6-square does not. The decomposition of an 8- or 16-square into components corresponds to octal (base 8) or hexadecimal (base 16) notation, alternatives to our chauvinistic ten-fingered decimal system. Even more structure is visible if these examples are rewritten in binary (base 2) notation, as was explained by Christopher J. Henrich in 1991. (He described the famous 8- and 16-squares, but his ideas extend to the others as well.[18]) It is worth observing that octal and hexadecimal numbers can be converted to binary very easily, since their respective bases (8 and 16) are both powers of 2.

This wealth of mathematical order was not enough for Franklin, who once wrote to James Logan in a moment of regret: "The magical squares, how wonderful soever they may seem, are what I cannot value myself upon, but am rather ashamed to have it known I spent any part of my time in an employment that cannot possibly be of any use to myself or others."[19] Personally, I'd like to think that this quotation was the invention of Jared Sparks, an early editor of Franklin's writings, since the original source is now lost and this passage is remembered only through Sparks's compilation. A later editor claimed that while the aptly named Sparks "preserved from oblivion many historical papers . . . he was disloyal to his author, and took liberties with his documents. He corrected and altered at pleasure."[20] But no one has ever alleged that any of those documents was an actual fabrication, and I suppose we shall have to accept that the harsh self-judgment printed by Sparks comes from Franklin's own pen. Yet it is only one more example of false humility, his words belied by his continued indulgence in this guilty pleasure for many years after 1750.

Is the magic square truly devoid of useful application? As a motif in literature, music, and the visual arts it still finds expression today. Mystics of many traditions tout its supposed powers. Still, in each of these examples, some other motif could substitute equally well. A more concrete application was suggested by the seventeenth-century Jesuit scientist Athanasius Kircher, who proposed a cipher based on the Jupiter magic square. He disassembled a phrase into syllables and scrambled them in a matrix according to the numerical positions in the square. (With only sixteen syllables to work with, it was not terribly secure.[21]) A well-known if far-fetched numerological analysis of *Moby Dick* claims that Herman Melville used a cipher based on the square in Dürer's *Melencolia I*.[22] One of Franklin's French contemporaries promoted a system that combined a 7×7 magic square with a sort of early Morse code.[23] Again, though, another matrix would serve just as well as cryptographic key; even today, the use of magic squares (and even Franklin magic squares) in encryption appears to confer no special advantage over other methods. If anything, these examples illustrate the magic square's perennial hold on human imagination, but no indispensable purpose beyond aesthetics and entertainment.

No, to find a true application of magic squares we must return to the special case of *Latin* squares. In the middle of the last century, Claude Shannon and other founders of information theory used Latin squares to construct codes that reduce the possibility of garbled transmissions. Latin squares and related concepts are still used today in modern coding theory, cryptography, combinatorics, and statistics.[24] They play a central role in the field of experimental design. To take the simplest case, say you want to test the effects of three pesticide treatments that protect crops in three different climates. The table below shows how you might conduct trials on three different crops *A*, *B*, and *C* using a Latin square design:

	Climate 1	Climate 2	Climate 3
Pesticide 1	*A*	*B*	*C*
Pesticide 2	*B*	*C*	*A*
Pesticide 3	*C*	*A*	*B*

Every pesticide-climate pairing appears at least once, as does every pesticide-crop pairing and every crop-climate pairing. This is accomplished in only nine trials, whereas it would take twenty-seven trials to include every pesticide-climate-crop triplet. More variables can be accounted for if we use Greco-Latin squares or even more complex sets of "mutually orthogonal" Latin squares. There are variations on the concepts of *orthogonal*, *Latin*, and even *square* that are likewise useful. And there is no need to restrict ourselves to agricultural examples; the same abstract design applies to medical research on human beings and to other areas of empirical scientific inquiry. To the extent that la Hire and Franklin and Euler founded the science of orthogonal magic squares, they deserve at least partial credit for these standard tricks from the modern experimenter's toolbox.[25]

Who can say whether some other, future use might arise from what began as a purely conceptual notion, originally designed only for divination or amusement. After all, ancient parlor tricks with amber were "useless" for millennia, before electricity became an indispensable part of modern life. Or, consider the Montgolfier brothers' hot air balloons, a landmark invention in the history of flight. Franklin witnessed a demonstration, and was asked by another member of the crowd in attendance: of what use is it? To this he replied: of what use is a newborn baby?

Viewed in this new light, it becomes apparent that his mathematical life, while not an especial source of pride for him personally, was another feather in an already well-plumaged cap. It is clear that, despite the leanest early preparation, Franklin managed to attain a small but secure place in the history of mathematics.

Notes

1. " 'Magic Square' Arrangement of Signatures in Declaration of Independence Points to Franklin," ["BF as author of United States Declaration of Independence"], on file at The Historical Society of Pennsylvania. Includes a press release from June 25, 1986, together with newspaper clippings and a

reference list of articles that appeared throughout the United States and Canada between 1981 and 1984.

2. Derrick N. Lehmer, "Les carrés magiques de Franklin," *L'Enseignement Mathèmatique*, Vol. 37, 1938, pp. 302-17; also the brief announcement "On Franklin Magic Squares," *Bulletin of the American Mathematical Society*, Vol. 36, 1930, p. 60.

3. Maya Mohsin Ahmed, "How Many Squares Are There, Mr. Franklin?: Constructing and Enumerating Franklin Squares," *American Mathematical Monthly*, Vol. 111, 2004, pp. 394–410.

4. Letters from Jacques Bouteloup to the author, July 24 and 31, 2002. Bouteloup is the author of *Calcul matriciel* (1961), *L'algèbre linéaire* (1967), and, most germane to the present discussion, *Carrés Magiques, Carrés Latins et Eulériens* (1991). He also authored "Vagues, marées, courants marins" [Marine waves, tides, currents], more common ground for an imaginary conversation with Franklin (who mapped the Gulf Stream and experimented with a method of calming turbulent waves).

5. Personal communications with Ray Hagstrom (Feb. 1–June 7, 2004) and Peter Loly (May 13–19, 2005).

6. Daniel Schindel, Matthew Rempel, and Peter Loly, "Enumerating the bent diagonal squares of Dr Benjamin Franklin F.R.S.," *Proceedings of the Royal Society A: Physical, Mathematical and Engineering*, Vol. 462, 2006, pp. 2271–2279.

7. Ray Hagstrom, "How Magic Can Squares Be?" Preprint, 2004.

8. Some others, both amateur and professional, who bear mention are Peter Bartsch (Germany), Miguel Angel Amela (Argentina), W. G. Ziegler (Canada), Lì Li (China), and Arno van den Essen (Radboud University, The Netherlands).

9. James P. Warbasse, Jr., Letter, *Pennsylvania Gazette*, Vol. 89, Dec. 1990, pp. 10–12.

10. Clarence C. Marder, *The Intrinsic Harmony of Number, Vol.1: The Magic Squares of Benjamin Franklin*, New York: Brick Row Book Shop, 1940. Biographical sketch is taken from a variety of public sources, mainly the U.S. Census (1880–1930).

11. Claude Bragdon, "Man: The Magic Square," in *The New Image*, New York: Alfred A. Knopf, 1928, p. 165. The magic elevator in figure 9.1 is taken from Claude Bragdon, "The Problem of Ornament," *Architecture*, Vol. 65, pp. 205–208 (April 1932).

12. *The Original Sudoku Book 2*, New York: Workman Publishing Company and Nikoli Co., 2005, pp. xiii–xiv.

13. This quote is from "The Course of a Craze," graphic to accompany "A Few Words About Sudoku, Which Has None," *New York Times (Week in Review)*, Aug. 28, 2005. See also Adam Fifield, "This little puzzle sure has our number," *Philadelphia Inquirer*, Dec. 1, 2005, p. E1.

14. Poignard, 1704 (see endnote 30, chapter 2); and Hutton's *Dictionary*, 1795 (see chapter 6), in an entry that essentially recapitulates material from *The Diarian Repository*, 1774.

15. "Recherches sur une nouvelle espece de quarres magiques," 1782 (read to the St. Petersburg Academy, 1779). See J. Dénes and A. D. Keedwell, *Latin Squares and their Applications*, New York and London: Academic Press, 1974; also Damaraju Raghavarao, *Constructions and Combinatorial Problems in Design of Experiments*, New York: Dover, 1988.

16. I suspect that Euler was thinking of numerical values, not letters, all along. See "De Quadratis Magicis," 1776. A remarkably similar method was discovered independently in Japan only five years later (Mikami, 1913, p. 292).

17. If you only look at the first two rows, it appears that the quotient square may be Latin. It isn't.

18. Christopher J. Henrich, "Magic Squares and Linear Algebra," *American Mathematical Monthly*, Vol. 98, 1991, pp. 481–488. The 8-square in chapter 7 and the 16-square in chapter 8 follow a similar decomposition, as Donald Knuth has pointed out (letters of Aug. 11, 2001 and Sept. 18, 2003).

19. Jared Sparks, ed., *The Works of Benjamin Franklin; containing several political and historical tracts not included in any former edition, and many letters, official and private, not hitherto published, with notes and a life of the author*, Boston: Hilliard, Gray, & Co., 1836–1840.

20. Smyth, *Writings*, vol. 1, p. i.

21. George E. McCracken, "Athanasius Kircher's Universal Polygraphy," *Isis*, Vol. 39, No. 4, 1948, pp. 215–228. Not coincidentally, Kircher was a follower of Trithemius, who was Agrippa's source for the planetary squares.

22. Viola Sachs, "The Gnosis of Hawthorne and Melville: An Interpretation of *The Scarlet Letter* and *Moby Dick*," *American Quarterly*, Vol. 32, No. 2, 1980, pp. 123–143.

23. "An Essay Tending to Improve Intelligible Signals . . . Translated From the French," *Transactions of the American Philosophical Society*, Vol. 4, 1799, pp. 162–173; read before the Society, 20 June 1788.

24. J. Dénes and A. D. Keedwell, *Latin Squares: New Developments in the Theory and Applications* (Annals of Discrete Mathematics 46), Amsterdam: Elsevier Science Publishers, 1991; and A. S. Hedayat, N.J.A. Sloane, and John Stufken, *Orthogonal Arrays: Theory And Applications*, New York: Springer, 1999.

25. Dénes and Keedwell, 1974. Also Kenneth P. Bogart, *Introductory Combinatorics*, 3rd edition, San Diego: Harcourt/Academic Press, 2000.

Acknowledgments

everal brief passages are adapted from my article "The Lost Squares of Dr. Franklin: Ben Franklin's Missing Squares and the Secret of the Magic Circle" (*American Mathematical Monthly,* Vol. 108, June–July 2001), used by permission of the Mathematical Association of America. With that exception, the research presented here has not been published previously. Parts of chapter 9 were included in my 2002 talk "The Followers of Franklin," presented at a section meeting of the American Mathematical Society but never published. Some of the material in chapter 4 regarding Franklin and education was included in the opening remarks to "Recreational Mathematics: A Short Course in Honor of the 300th Birthday of Benjamin Franklin" (MathFest 2005).

The research that became this book would not have been possible without the existence of many outstanding libraries and their appointed guardians. I would like to thank the following individuals for their expert assistance: James N. Green and Connie King, Library Company of Philadelphia; Jill Lee, Athenæum of Philadelphia; Lynne Farrington and Doane Hollins, University of Pennsylvania; Elizabeth Fuller and Karen Schoenewaldt, Rosenbach Museum and Library; Glenys A. Waldman, Masonic Library and Museum of Philadelphia; R. A. Friedman and Max Moeller, Historical Society of Pennsylvania; Jonathan Stayer, Pennsylvania State Archives; Barbara Bair and

Patrick Kerwin, Library of Congress Manuscript Division; Jessica Murphy, Dibner Institute for the History of Science and Technology, MIT; Sr. Martha Jacob, Ursuline Sisters of Louisville; Christine Woollett and Gudrun Richardson, The Royal Society; Sally Brooks, Museum of London; and Stephen Freeth, Guildhall Library Manuscripts Section. By far the greatest debt is owed to the people at the American Philosophical Society Library: Valerie-Anne Lutz, Charles B. Greifenstein, Robert S. Cox, and most of all Roy G. Goodman, Curator of Printed Materials, who knows Franklin better than most people know themselves.

I am grateful to Victor Katz (University of the District of Columbia) and V. Frederick Rickey (United States Military Academy), whose Institute in the History of Mathematics and Its Use in Teaching first inspired me to pursue my own historical research. For ongoing encouragement throughout the life of this project, I must thank David Zitarelli (Temple University) and Robert E. Bradley (Adelphi University). Thanks are also owed to Radcliffe Edmonds (Bryn Mawr College) for resolving several linguistic issues; to my colleagues in the Department of Mathematical Sciences at Villanova University for providing a hospitable environment for unorthodox research; to Stephen Hague, Executive Director at Stenton; to Ellen R. Cohn and Kate M. Ohno, Editor and Assistant Editor respectively, of *The Papers of Benjamin Franklin*; to Brigitte Comparini of the Packard Humanities Institute, for allowing me to consult a pre-production version of the digitized *Papers* for research purposes; to Joseph Dworetzky, for legal advice; and to my father Aris Pasles, a former engineer, who first taught me mathematics. The manuscript was read by my wife Elise Burns Pasles (Bryn Mawr College), my brother George Pasles (wandering mathematician of no fixed affiliation), Kathleen Ambruso Acker (Cabrini College), Brian J. Jacobs, Professor Katz, and Professor Zitarelli, all of whom offered invaluable suggestions. Elise tolerated my involvement in this project for far longer than was reasonable. (Anyone interested in generalizations of orthogonal Latin squares would do well to consult her Ph.D. dissertation.) Finally, my heartfelt gratitude is extended to Vickie Kearn, editor extraordinaire; to Dimitri Karetnikov, illustration specialist; to Natalie Baan, production editor; and to everyone else at Princeton University Press.

Appendix

1. A BASIS

Every 4×4 Franklin magic square can be obtained by adding multiples of these seven simple matrices:

1	1	1	1
1	1	1	1
1	1	1	1
1	1	1	1

1	0	0	−1
0	0	0	0
0	0	0	0
−1	0	0	1

0	0	0	0
0	1	−1	0
0	−1	1	0
0	0	0	0

0	1	−1	0
0	0	0	0
0	0	0	0
0	−1	1	0

0	0	0	0
1	0	0	−1
−1	0	0	1
0	0	0	0

1	1	−1	−1
−1	−1	1	1
0	0	0	0
0	0	0	0

1	−1	0	0
1	−1	0	0
−1	1	0	0
−1	1	0	0

If we drop the last two of these matrices, we get a basis for the 4×4 *panfranklin* squares, that is, those in which all 16 bent rows sum magically. The latter species necessarily repeats some entries.

2. THE FRANKLIN-STRACHEY METHODS (for mathematicians)

Franklin's 6-square is built according to a general method that works for all even orders greater than 4. Let $n = 2k$, $k \geq 3$, and let A

be a $k \times k$ fully magic square (rows, columns, and two diagonals sum equally) with entries $1, 2, \ldots, k^2$. Define J_t to be the $t \times t$ constant matrix of ones. Put

$$M = J_2 \otimes A + k^2 \begin{bmatrix} 0 & 3 \\ 2 & 1 \end{bmatrix} \otimes J_k.$$

For those unfamiliar with the Kronecker product $\otimes$, this expression abbreviates the 6×6 procedure described in chapter 8, if we choose k to be 3 and A to be the lo shu matrix (chapter 2).

The matrix M has equal row sums, but not equal column sums. We amend this discrepancy as follows. Let (r, s) be any solution to the Diophantine equation $3r - s = k$ such that $r, s > 0$. Pick r rows in the upper half of the matrix and s rows in the lower half. In each of these chosen rows, exchange the first half-row with the second half-row. The resulting matrix M' now has constant row, column, and bent-row sums. Thus it is an $n \times n$ Franklin magic square. This method can be generalized further by allowing t to take on other values, but it is far more complicated.

The Kronecker product of fully magic (straight-diagonal) squares is a fully magic square. Since $6 = 2 \times 3$, and there is no magic square of order 2, it would appear that an insurmountable obstacle stands in the way of using the product in this case. The brilliance of the Franklin-Strachey methods is in resolving this quandary creatively. (For the record, the product of two Franklin magic squares is not usually Franklin magic. However, the product of a Franklin magic with a fully magic square—taken in that order—*is* Franklin magic. This leads to many more ways to construct Franklin-style squares.)

The statistics in chapter 8 (describing the number of ways the Franklin method may be applied) are justified as follows. Since the number of fully magic squares possible in a given order k remains unknown, let us use the very conservative (but correct) estimate that there is at least one such square possible for any size. For given k, the value of r ranges from $\lceil k/3 \rceil$ to $\lfloor 2k/3 \rfloor$, while $s = 3r - k$. The number of ways to choose r rows from the upper half of the matrix and *s* rows from the lower half is $\binom{k}{r}\binom{k}{s}$. Thus the number of ways to apply Franklin's method is at least

$$\sum_{r=\lceil k/3\rceil}^{\lfloor 2k/3\rfloor}\binom{k}{r}\binom{k}{3r-k},$$

which can be expressed in closed form using hypergeometric functions. Incidentally, this same expression also appears in a completely different context, enumerating random walks that are subject to certain constraints.

3. A MAGIC SQUARE OF PHILIPPE DE LA HIRE

The *Diarian Repository* (1774) gives an example of two 8×8 component squares that bear striking similarity to the three sets of components described for Franklin's squares. (Strictly speaking, la Hire uses 1,2, . . . ,8 in his remainder square. We have subtracted 1 from every entry for consistency with the presentation in Chapter 9.)

Quotients

3	4	3	4	4	3	4	3
0	7	0	7	7	0	7	0
2	5	2	5	5	2	5	2
6	1	6	1	1	6	1	6
1	6	1	6	6	1	6	1
5	2	5	2	2	5	2	5
7	0	7	0	0	7	0	7
4	3	4	3	3	4	3	4

Remainders

2	4	0	1	6	7	3	5
5	3	7	6	1	0	4	2
2	4	0	1	6	7	3	5
5	3	7	6	1	0	4	2
5	3	7	6	1	0	4	2
2	4	0	1	6	7	3	5
5	3	7	6	1	0	4	2
2	4	0	1	6	7	3	5

4. A COMPARISON OF TWO "NEW" SQUARES

There is a remarkable connection between the 8-square Franklin sent to Dubourg in 1772 and the 16-square that appears in the Canton Papers for 1765 (Chapter 8). Represent the size of each matrix by n. In both, we fill the top two rows with the entries 1, 2, . . . , n and their complements. Moreover, the 2×1 submatrices:

n
1

2
$n-1$

$n-2$
3

(and so on) appear within those two rows. The pattern can be continued throughout both magic squares.

While these two magic squares are filled in row by row, the more famous 8- and 16-squares are filled in column by column; but this is an immaterial difference. (That the columns are not always contiguous is likewise of little consequence.) In both cases, the row or column pairs are filled with 1,2, . . . ,n and their complements, then $n+1$, $n+2$, . . . , $2n$ and their complements, and so on. There the similarities end. In the famous squares, within each column pair, every value is paired with its complement; but in these obscure examples, a more complex relationship is apparent (1 paired with n, and 2 with $n - 1$, for example). Thus the famous squares are somewhat different from the ones described in this note.

5. Solutions to Boxed Problems

Chapter 2, page 50

Flip Agrippa's Jupiter square horizontally (left-to-right), so that the order of the columns has been reversed. Its first row now reads 1, 15, 14, 4. Compare this to the *Ladies Diary* square with each entry reduced by seven. You'll notice some similarities: the smallest values (1, 2, 3, 4) are identically placed in both magic squares, the second-smallest values (5, 6, 7, 8 vs. 7, 8, 9, 10) are also identically placed, and so on. If you subtract that (flipped) Agrippa square from the (reduced) *Diary* square, you get a magic square of zeroes, twos, fours, and sixes, which can be built using M and its rotations and reflections. In short:

$$\text{Diary square} = \text{``flipped Agrippa''} + \text{matrix of sevens} + 4\,M$$
$$+ 4\,M^H + 2\,M^T + 2\,M^R,$$

where M^H is obtained by flipping M horizontally, M^T by flipping M along the diagonal that joins the northwest and southeast corners, and M^R by rotating M clockwise 90 degrees.

Chapter 3, page 69

$1 + 2 + 3 + \;\ldots\; + 12 = (1 + 12) + (2 + 11) + \;\ldots\; + (6 + 7) = 13 + 13 + 13 + \;\ldots\; + 13 = 6 \times 13 = 78$ guns. Incidentally, he expects "a great saving of powder in cloudy days" (*Poor Richard improved for 1757*).

A general formula for adding the first n counting numbers, derived in like fashion, is $S = n\,(n + 1)/2$.

Chapter 4, pages 90–91

These exercises are taken from the 1741–1749 issues of the *Diary*.

a) No. 61, 1747. Let x be such a number. We know that $x + 1$ is evenly divisible by 2, 3, 4, 5, ..., 20. Thus in particular it is divisible by the primes 2, 3, 5, 7, 11, 13, 17, 19; and some of these must appear multiple times, in order to accommodate 4, 6, 8, 9, and so on. The smallest such number is $2^4 \cdot 3^2 \cdot 5 \cdot 7 \cdot 11 \cdot 13 \cdot 17 \cdot 19$, and doubling or tripling it will produce the next two such numbers. Subtract 1 from each of these to get the final answer: 232,792,559, 465,585,119, and 698,377,679.

b) No. 4, 1741. The volume can be maximized using the standard optimization methods of calculus, after gathering the necessary ingredients by means of the Pythagorean Theorem. If the term "slant side" refers to an edge length, then the pyramid has height $20\sqrt{3}$ and a square base of side $40\sqrt{3}$; it is half as high as it is wide. The "solidity" (volume) is $32{,}000\sqrt{3}$, or around 55,425.63 cubic inches. If, on the other hand, the "slant side" refers to the altitude of one face (this is more commonly called the *apothem* or *slant height*), then the pyramid has height $20\sqrt{3}$ and a square base of side $40\sqrt{6}$, and the volume is $64{,}000\sqrt{3} \approx 110{,}851.25$ cubic inches. Either way, the height of the pyramid is the same!

c) No. 54, 1746. ("Right lines" are perpendicular to the sides.) The meadow has sides 58.755, 62.611, and 65.697 chains, and area 1,673 square chains.

d) No. 6, 1741. Furlongs can be converted to feet, and acres to square feet, courtesy of Poor Richard's "Profitable Notes" (previous chapter). The length, breadth and depth of a stone are 12, 6, and 4 inches, respectively. Approximately 244,650 stones would be sufficient to wall the circular arc, depending on precisely how the stone-layer chooses to reconcile flat stones with curved wall. Incidentally, the shape of Mr. Heath's park is taken from the *Ladies Diary*, 1741.

e) No. 58, 1746. $396 \pm 66\sqrt{6}$ feet, $132\sqrt{21}$ feet. As decimals: 234.33, 557.67, 604.90.

f) No. 2, 1742. $\frac{100\pi}{51+10\sqrt{2}} \approx 4.822673718$ feet. (The original question reads: "There is a Quadrant of a Circle, whose Radius is $= 10$ Feet, and in it there is a Circle inscrib'd, which touches both the Radii, and the Concavity of the Limb thereof; and, in one of the Curve Corners, there is a little Circle inscrib'd; so that it touches both the Radius of the Quadrant, the Concavity of the Limb, and the Convexity of the in-scrib'd Circle; The Area of this little Circle is requir'd?")

Chapter 4, page 98

Answer: The privateer overtakes the ship in 5 hours, 42 minutes, $51\frac{3}{7}$ seconds. (A similar problem can be found in James Stewart, *Single Variable Calculus*, fifth edition, Brooks/Cole, 2003, p. 684, no. 8.) Our answer assumes that the craft is so maneuverable that its captain may instantaneously change course in response to the pursued ship, and that the latter never deviates from a straight course upon detecting the pursuit.

Technically a privateer differs slightly from a pirate: he enjoys state sanction (though this is unlikely to console his victim).

Chapter 8, page 203

(a) As before, you can pair up the first and last values $(1 + 256)$, then the second and second-to-last, and so on. Each pair sums to

257, and there are 128 pairs (the last one being $128 + 129$). Therefore the sum of the first 256 positive integers is $128 \times 257 = 32{,}896$. This is the sum of all entries in the square. (Notice that the argument relies on the fact that changing the order of a list of numbers does not change their sum, and neither does associating them in pairs in this way. Addition is both *commutative* and *associative*.) Since there are 16 rows which sum together to 32,896, each row alone sums to $32{,}896/16 = 2056$. (b) By the same reasoning, the magic sum is $(n^2/2)(n^2 + 1)/n = n(n^2 + 1)/2$.

Chapter 8, page 206

(a) 3 years, 305 days, 11 hours, 6 minutes, 40 seconds. (b) No, because a 28-square has sum 10,990 and a 29-square has sum 12,209. The key is that the formula assumes the entries to be $1, 2, \ldots, n^2$. Dubourg's square must use some other, less traditional, set of numbers.

Chapter 8, page 210

(a) Many possible solutions. For example, *quadruple* each entry and add 1245. (b) He did not have to begin with an 8×8 magic square. To take just one example, he might use a 4×4 magic square, double each entry and add 2733.

Further challenge for anyone who's ever taken a course in number theory: Find all such strategies for $n = 8$. That is: Find all integers a, b for which the linear transformation $x \mapsto ax + b$ turns a magic square with entries $1, 2, \ldots, 64$ into a magic square with sum equal to 11,000. (Research question: what if n is arbitrary?)

Chapter 8, page 215

(a) 26; (b) $26! = 26 \times 25 \times 24 \times \ldots 3 \times 2 \times 1$. The cryptogram reads "The absent are never without fault, nor the present without excuse" (from *Poor Richard*, 1736).

Chapter 9, page 227

(a) He rounded down to £131,000. (More precisely, £1,000 × $1.05^{100} \approx$ £131,501.26.) (b) £31,501.26 × $1.05^{100} \approx$ £4,142,455.

Chapter 9, page 229

66 years old: $66 + 33 + 22 + 3 \times 3 = 130$.

Index